Lawer

Penguin Education

Sociology of Science

Edited by Barry Barnes

Penguin Modern Sociology Readings

General Editor

Tom Burns

Sociology of Science

Selected Readings

Edited by Barry Barnes

Penguin Books

Penguin Books Ltd, Harmondsworth,
Middlesex, England
Penguin Books Inc., 7110 Ambassador Road,
Baltimore, Md 21207, U.S.A.
Penguin Books Australia Ltd,
Ringwood, Victoria, Australia

First published 1972

Made and printed in Great Britain by
Cox & Wyman Ltd, London, Reading and Fakenham
Set in Monotype Times

Contents

Introduction

Today, after three centuries of exponential growth,[1] science exists as a major institutional complex in all modern societies; its cultural and economic significance is now universally recognized and indeed, for some, it is the key to an understanding of the characteristic features of 'advanced' capitalism. Yet sociologists have scarcely begun to examine its internal structure, and systematic investigation of its relationship with the wider society remains a hope for the future.

Part of the reason for this neglect doubtless resides in the low intelligibility of the beliefs and practices of science, that is, of its culture. The sociologist is likely to find this obscure and inaccessible, and accordingly to be discouraged from studying it. But that this cannot be the whole story is clear, when the prominence given by contemporary investigators to the sociology of the professions is considered, for similar problems of intelligibility occur in this area. Perhaps in taking its ideals as their own sociologists have been deterred from investigating science as a social institution or system of social action.

Whatever the reason, the classic writers of the sociological tradition left no detailed analyses of science from this viewpoint, and it was long before it became the subject of a differentiated research programme. This eventually occurred in the United States, where a build-up of interest during the 1930s resulted, after a period of stagnation, in the firm establishment of the sociology of science as a separate academic specialty.

A dominant influence in this development was the work of Robert Merton, both as writer and teacher. By 1945 Merton had laid down an approach which identified science as a social institu-

1. Price (1963), using a number of quantitative indicators, has demonstrated that, since the appearance of its first periodicals before 1700, science has been doubling in size every ten to fifteen years. An unrestricted exponential increase of this sort is, he suggests, best explained by the hypothesis that internal development has been the main cause of scientific growth, and that this has been little affected, in its general trend, by either external stimuli or economic and social restrictions.

tion with a characteristic ethos, and subjected it to functional analysis. This was for a long period the only theoretical approach available to sociologists in the area, and it remains productive and influential today. Its central ideas have received detailed elaboration, modification and reinterpretation by, among others, Barber, Hagstrom, Storer and Merton himself, making it the only maturely developed framework for the sociological study of science.[2]

The main achievement of this functionalist school has been to throw light on the process of internal social control in science, and to show how it depends on the allocation of internal recognition and honorific reward. Some work on the functional relationship of science to the wider society, and its interaction with other institutions, has also been stimulated. But the work of the school has, in practice, been limited in scope; problems related to culture have been increasingly ignored. As Downey (1967) has pointed out, no studies have been made of the general cultural impact of science. And the culture of science, its characteristic complex of meanings, where it has been analysed at all, has been treated either extremely superficially, or so gingerly as to hinder good scholarship.[3]

Only in the last decade has the monopoly of the approach weakened. Its theoretical assumptions have been questioned from a number of different standpoints in papers by Mulkay (see Part Two), Downey (1969) and Barnes and Dolby (1970). At the same time alternative theoretical perspectives have appeared and interest in cultural questions has increased, so that now the field is broader in scope but less theoretically unified than hitherto. It is this unsettled but essentially promising state that the readings of this volume try to reflect.

The two selections of the first Part, although chosen strictly for

2. Although functionalism has been the dominant theoretical impetus to post war sociological studies, rather more work has been guided by purely practical concerns: how to maximise the productivity of scientists; how to achieve the most effective kind of laboratory organisation; how to establish effective criteria for the allocation of research funds; and so on. Kaplan (1964) has produced a review giving prominence to this kind of approach, and effectively criticising the functionalist analysis. But, at the same time, he stresses the limited results of this empirically oriented work and the need for more academic study of the sociology of science.

3. An admirable study of *serendipity* in science by Barber and Fox (1958) is the exception that proves the rule here. It clearly reveals the value of studying the concrete beliefs and practices involved in research.

their sociological interest, serve as a prologue drawing attention to the two most important periods of fundamental change in the history of science – the cultural revolution of the sixteenth and seventeenth centuries, and the social reorganisation of the nineteenth.

Although the first of these periods has been studied in detail by historians and some sociologists, the second has only recently claimed attention and offers a promising field for future work. The provision of fully differentiated scientific roles as careers, and of prolonged full-time training for them, their isolation and concentration into laboratories, and their formal division into disciplines and specialties, were all nineteenth-century innovations, which reflected the general trend towards professionalisation and specialisation characteristic of industrial society as a whole. Science was becoming organised into forms appropriate as the knowledge institutions of highly differentiated societies.

Part Two deals with internal features of pure or academic science. Functionalist approaches are included here, with particular stress being laid on their treatment of problems of social structure and organisation. But great prominence is also given to the work of Thomas Kuhn which provides insight into stability and change in the culture of science, and is currently being used by sociologists in this field to construct an entirely new theoretical approach to it. Kuhn's description of science (which this writer regards as a landmark in the process of understanding science as it is, rather than as it 'ought' to be) points the way to a new and deeper appreciation of its internal processes. At the same time it reveals, by contrast, how much the existing sociological analysis of science depends upon a preconceived empiricist model of scientific activity.[4]

4. Indeed an empiricist conception of natural science has affected social science in general, not just by shaping its methodological ideals, but also by influencing its theories and forms of explanation. This can be clearly seen in anthropology, as I have tried to show elsewhere (see Barnes, 1971); but it is also apparent in sociology, particularly in the structure of those general theories which make a fundamental distinction between 'rational' or 'instrumental' action, and other kinds. The credibility of these theories is likely to depend upon how successfully natural science can continue to provide some kind of concrete paradigm of empirical belief and instrumental action. In this sense the study of scientific activity, via its epistemological implications, could be of enormous consequence to sociology as a whole.

The increasing technological and economic relevance of science is considered in the third Part, together with the feed-back effects this has had upon science itself. There is a marked contrast here with the analyses in Part Two, which treat science as essentially isolated and autonomous, but that the two approaches are fully reconcilable will be readily apparent from the contributions of Price and Ben-David.

'Pure' and 'utilitarian' conceptions of science are, however, frequently perceived as conflicting; indeed a tension between the two has always been apparent in the writings of scientists themselves. This tension intensified during the nineteenth century when scientists, like other 'professionals', faced the problem of achieving and maintaining autonomy and status. On the one hand, a search for institutional and disciplinary autonomy encouraged dissociation from social demands and development of internal social control; narrowing and purification in the aims and methods of science were accelerated (as was, in consequence, the tendency of scientific vocabulary to become ever more specific and esoteric). These changes were carried farthest in the German Universities, where disinterested research won such admiration that its organisation and ideology came to be widely imitated. On the other hand scientists needed status, sufficient social recognition to ensure the support of their activity. Hence they were led, in some contexts, to stress its practical and economic value. A few went further than this, and attempted to extend the cultural influence of science in the wider society, necessarily at the expense of other accredited sources of knowledge. In such endeavours they stressed the power of the general methods of science, their rational, logical and objective nature and their wide applicability, as well as their utility.

The maintenance of autonomy and status has continued to engender essentially conflicting demands to the present time, and their practical and ideological reconciliation has been a continuing problem for scientists.[5] Moreover the opposed images

5. The well-known controversy in Britain exemplified by the contrasting writings of Bernal (1939 and 1949) and Polanyi (1951) although ostensibly concerned with the autonomy of science, can, with hindsight, be better interpreted as a conflict about how to effect this reconciliation; for the utilitarian Bernal was, in practice, greatly concerned with the maintenance

elaborated and popularised during the professionalisation period – aloof, mysterious and encapsulated on the one hand, utilitarian and iconoclastic on the other – have grown to structure the current social perception of science.

Nor, in this context, are they at all inappropriate or contradictory. The utility of the sciences has become manifest through technology; whole industries are based on their discoveries, and our entire stock of artefacts mutely proclaims their efficacy. Never has science attracted greater concern as a potential source of social change, a force capable of revolutionising whole areas of experience. Yet, at the same time, science, by its own internal development, has become ever more specialised, esoteric and inaccessible; in many ways the image of the ivory tower is more appropriate than ever before. Research remains a mysterious activity in the eyes of the general public, and increasingly is becoming intelligible only within the boundaries of a single scientific specialty.

Eventually sociologists must turn to a systematic and detailed examination of the relationship of science to the wider society; existing conceptions of this relationship must be tested and elaborated. At present, however, the study of this area is only beginning, and the existing images of science, rather than being treated as hypotheses, are providing investigators with alternative basic assumptions concerning the scope and potential influence of science and its methods. Their opposition thus underlies many current controversies in this area.

Basic to these controversies are opposed analyses of the nature of scientific activity itself. Is it a powerful, highly general method of investigating reality, guided by highly general norms for the evaluation of results; or is it intelligible entirely in terms of esoteric techniques, skills and theoretical structures, perfected within particular traditions of research? These alternatives imply different general formulations of the relationship of science and society. Do the sciences act as the source of a rationality increasingly

---

of scientific autonomy, whereas Polanyi justified autonomy by arguments including utilitarian criteria.

Daniel Greenberg (1969, especially ch. 6) has recently described how the United States scientific community faced a similar problem, in an acute and concrete form, at the end of the Second World War.

pervading the wider society; or are they isolated and encapsulated sub-cultures defined and unified only by Kuhnian paradigms? Do the theories of science offer a characteristic and total world view; or are they no more than pragmatic aids to particular types of concrete problems, losing all significance when abstracted from their practical context? Does the credibility of scientific pronouncements reside in the logic of their supporting argument and their power of prediction; or simply in the institutionalised authority of science?

In the absence of consensus on questions of this magnitude, it is not surprising that opinion is divided upon what specific effects science is producing in the wider society. Thus, some give science a major causal role in the general social processes of secularisation and rationalisation, claiming that its values are increasingly being established as those of industrial society as a whole; but it is equally plausible to argue that the general values of industrial societies, and those of their knowledge institutions, have shifted alike under the common influence of differentiation.

This last controversy points to a third possible sociological orientation to science. Instead of analysing it as an agent of social change, or as an encapsulated sub-culture, the extent to which its own social structure and culture are derivative of those of the wider society can be investigated. Surprisingly, this viewpoint, so characteristically sociological, has guided little research in recent years, despite having been applied with outstanding success in a number of earlier contributions.

The three concluding Parts of the book illustrate all these types of approach, with a strong emphasis on cultural questions. Examples of social change produced entirely through the medium of technology have been omitted; they are better treated at greater length in other contexts. But I have otherwise sought to cover a variety of approaches to the relationship of science and society, and have stressed no single type of analysis.[6]

6. I have lacked space to cover some recent exciting anthropological work, potentially of great significance. The reader is referred to Robin Horton's admirable 'African traditional thought and Western science' (1967) and Mary Douglas's two intensely stimulating essays on the relationship of world views and social structure, 'Purity and Danger' (1966) and 'Natural Symbols' (1970).

The overall selection, then, reflects a broad conception of the scope of the sociology of science, and attempts to meet a wide range of requirements: it is designed to reveal the possibilities of different kinds of sociological theory within the field, to reflect different preconceptions of the general relationship between science and the wider society, and to balance 'social structural' with 'cultural' analysis. Bearing in mind limitations of space, I have sought merely to convey important ideas and approaches, rather than to include work which presents large amounts of unstructured information, or justifies statements of restricted interest by techniques of great methodological sophistication. Despite this policy, and the comparatively small size of the research literature, I have, of course, been compelled to omit many essential contributions and neglect significant areas of research. Hence, I must end by stressing that, overriding all specific aims, the dominant concern of these Readings has been to generate interest, and stimulate further reading in their subject.

References

BARBER, B., and FOX, R. C. (1958), 'The case of the floppy-eared rabbits: an instance of serendipity gained and serendipity lost', *Amer. J. Sociol.*, no. 64, pp. 128–36.

BARNES, S. B., and DOLBY, R. G. A. (1970), 'The scientific ethos: a deviant viewpoint', *European Sociol.*, vol. 11, pp. 3–25.

BARNES, S. B. (1971), 'The comparison of belief systems: anomaly versus falsehood', in R. Finnegan and R. Horton (eds.), *Modes of Thought*, Faber.

BERNAL, J. D. (1939), *The Social Function of Science*, Routledge & Kegan Paul.

BERNAL, J. D. (1949), *The Freedom of Necessity*, Routledge & Kegan Paul.

DOUGLAS, M. (1966), *Purity and Danger*, Routledge & Kegan Paul; Penguin, 1970.

DOUGLAS, M. (1970), *Natural Symbols*, Cresset.

DOWNEY, K. J. (1967), 'Sociology and the modern scientific revolution', *Sociol. Q.*, vol. 8, pp. 239–54.

DOWNEY, K. J. (1969), 'The scientific community: organic or mechanical', *Sociol. Q.*, vol. 10, no. 4, pp. 438–54.

GREENBERG, D. (1969), *The Politics of American Science*, Penguin.

HORTON, R. (1967), 'African traditional thought and Western science', *Africa*, no. 37, pp. 50–71, and 155–87. (An abridged version appears in *Witchcraft and Sorcery*, Penguin, 1970.)

KAPLAN, N. (1964), 'Sociology of science', in R. E. L. Fanis (ed.), *Handbook of Modern Sociology*, Rand McNally.

POLANYI, M. (1951), *The Logic of Liberty: Reflections and Rejoinders*, Routledge.

PRICE, D. J. de S. (1963), *Little Science: Big Science*, Columbia University Press.

Part One
The Emergence and Institutionalization of Modern Science

'What was it that happened in Renaissance Europe when mathematics and science joined in a combination qualitatively new and destined to transform the world?' Thus Joseph Needham, in the first essay of this section, presents an analysis of the nature of modern science and prepares to illuminate its origin through comparative method. If Needham is right the period between Copernicus and Newton saw beliefs being influenced by social structure in such a way as to transform qualitatively the nature of science, and thus lay down the form of much that counts as knowledge in the modern world. But, as his contribution makes clear, others stress different causal influences as central in the rise of modern science, and historians with little inclination to generalise are always ready to provide over-ambitious theories with numerous embarrassing counter-instances.

Not all historians stress the importance of the Renaissance as strongly as Needham. Many see the enormous conceptual changes of this period as the fruition of an earlier tradition, developing internally. But although the scientific achievement of the medieval period is now well established, and the myth of its intellectual stagnation dispelled, most scholars still locate the key events in the rise of modern science in the Europe of the sixteenth and seventeenth centuries. And this is the interpretation that all sociological studies have followed.

Among studies stressing the role of external social factors, familiar tensions are apparent between explanations in terms of dominant socio-economic patterns and those which point to the influence of beliefs and values. Engels' view (see Marx and Engels, 1951, vol. 2, p. 457) 'If society has a technical need, that helps science forward more than ten universities,' found its most single-minded application in Boris Hessen's 'The social and economic roots of Newton's *Principia*' (1931); this interprets Newton's work in the light of a strict economic determinism. When it was published in 1931 few were able to set aside their political commitments and evaluate it objectively, but it provided an influential theoretical model, and one may wonder how many of the empirical studies now used to illustrate its weaknesses would have existed in its absence. In contrast, Edgar Zilsel's work (1941, for example) which claims that capitalism assisted science by eroding the barriers between scholars and craftsmen and facilitating a union of practical and theoretical knowledge, has uniformly commanded the respect, if not the agreement, of all historians.

Robert Merton's study (1938) of science in seventeenth-century England maintains a fine sensitivity to economic factors, but here, stress is laid on the importance of religious beliefs and values; puritanism is held to have played an important part in generating and maintaining interest in science. Because of its methodical and quantitative treatment of historical materials, and its compatibility with Weber's work on the connection of protestantism and capitalism, this major study is widely admired among sociologists; unfortunately, few have felt competent to follow in its tracks, still less to test it, and one must instead thank historians for considering it worthy of criticism.

These accounts comprise but a small part of an historical literature, all of which is full of sociological interest. Much of this is the work of 'internalist' historians, which stresses intellectual factors in scientific growth and forms a valuable corrective to sociological macro-theorising. Internalist accounts are not however incompatible with sociological explanation as such; Thomas Kuhn's account (1957) of the Copernican Revolution illustrates this point, but it is true, logically, even of the work of such historians as Koyré and Hall. Moreover, such work is of more than just negative interest to the externalist – for he must demonstrate how his external causal factors modify the norms and beliefs of cultures or sub-cultures. External causes do not generate new activities *ab initio*. They have to be understood as they operate on the matrices of belief and action described by the intellectual historian.

Would that the institutional history of science thrived as its intellectual history! The institutional locale of scientific activity must surely always have been a major influence upon it. Certainly, the provision of fully differentiated scientific roles during the nineteenth century has been necessary to many subsequent intellectual developments. Joseph Ben-David (1965, 1966) has gone as far as to suggest that the existence of some type of scientific role, of social groups conceiving of themselves as scientists, is always crucial to any worthwhile scientific achievement. Whether or not this is a justified claim, much is owed to the few scholars, such as Ben-David, and Daniels (1967), who have studied institutional change in science, and recognised the particular importance of the professionalisation process that began a century and a half ago.

The second contribution here is taken from Ben-David's and Zloczower's comparative study of 'Universities and Academic Systems in Modern Societies'. It attempts to relate institutional change in the German Universities to features of the surrounding social structure, and to show how competition within the ensuing system of roles stimulated scientific progress.

References

BEN-DAVID, J. (1965), 'The scientific role: conditions of its establishment in Europe', *Minerva*, vol. 4, pp. 15–54.

BEN-DAVID, J., and COLLINS, R. (1966), 'Social factors in the origin of a new science: the case of psychology', *Amer. sociol. Rev.*, vol. 31, no. 4, pp. 451–65.

DANIELS, G. H. (1967), 'The process of professionalization in American Science: the emergent period, 1820–60', *Isis*, vol. 58, no. 192, pp. 151–67.

HESSEN, B. (1931), 'The social and economic roots of Newton's *Principia*', in *Science at the Crossroads*, Kniga. (A new edition is forthcoming from Cass.)

KUHN, T. S. (1957), *The Copernican Revolution*, Harvard University Press.

MARX, K., and ENGELS, F. (1951), *Selected Works*, Laurence & Wishart.

MERTON, R. K. (1938), 'Science, technology and society in seventeenth-century England', *Osiris*, vol. 4, pp. 360–62. (Reprinted by Howard Fertig, 1970.)

ZILSEL, E. (1941), 'The sociological roots of science', *Amer. Sociol.*, vol. 2, pp. 544–60.

1 Joseph Needham

Mathematics and Science in China and the West

J. Needham, 'Mathematics and science in China and the West', *Science and Society*, vol. 20, 1956, pp. 320–43.

The aim of the following pages is to open a discussion on the relations of mathematics to natural science in two diverse civilizations, Europe and China. In the preparation of a comprehensive work on the history of science, scientific thought and technology in the Chinese culture area, it became evident that the theme of the present paper was worth investigation. Exactly what were the relations of mathematics to science in ancient and medieval China? What was it that happened in Renaissance Europe when mathematics and science joined in a combination qualitatively new and destined to transform the world? And why did not this arise in any other part of the world? Such are some of the questions which arose.

Nature and achievements of Chinese mathematics

The first thing to do is to get the perspective right. No mathematical works before the Renaissance were at all comparable in achievement with the wealth and power of the developments which took place afterwards. It is pointless, therefore, to subject the old Chinese contributions to the yardstick of modern mathematics. We have to put ourselves in the position of those who had to take the earliest steps and try to realize how difficult it was *for them.* Measured in terms of human labor and intellectual attack, one can hardly say that the achievements of the writers of the *Chiu Chang Suan Shu* (first century BC) or of the initiators of the Thien Yuan algebra (thirteenth century) were less arduous than those of the men who opened new mathematical fields in the nineteenth century. The only comparison which can be made is between the old Chinese mathematics and the mathematics of other ancient peoples, Babylonian and Egyptian in their day, Indian and Arabic.

Chinese mathematics was quite comparable with the pre-Renaissance achievements of the other medieval peoples of the Old World. Greek mathematics was doubtless on a higher level, if only on account of its more abstract and systematic character, seen in Euclid; but, as we have noted, it was weak or tardy just where the mathematics of India and China (more faithfully based, perhaps, on those of the Babylonians) were strong, namely, in algebra.

'On account of its more abstract and systematic character' – so came the words of themselves to the keys of the machine. Systematic, yes, there no doubt is possible, but abstract – was that wholly an advantage? Historians of science are beginning to question whether the predilection of Greek science and mathematics for 'the abstract, the deductive and the pure, over the concrete, the empirical and the applied' was wholly a gain. According to Whitehead:

> It is a mistake to think that the Greeks discovered the elements of mathematics, and that we have added the advanced parts of the subject. The opposite is more nearly the case; they were interested in the higher parts of the subject and never discovered the elements. . . . Weierstrass' theory of limits and Cantor's theory of sets of points are much more allied to Greek modes of thought than our modern arithmetic, our modern theory of positive and negative numbers, our modern graphical representation of the functional relation, or our modern idea of the algebraic variable. Elementary mathematics is one of the most characteristic creations of modern thought – characteristic of it by virtue of the intimate way in which it correlates theory and practice.[1]

How far the mathematics of ancient and medieval China helped to lay the foundations for the more difficult, because most simple and elementary techniques of handling the concrete universe, we hope to make clear elsewhere. In the flight from practice into the realms of the pure intellect, Chinese mathematics did not participate.[2]

1. Whitehead (1948, pp. 132 et seq.). The essay of Farrington which comments upon this passage is well worth reading.

2. It is interesting that the converse process seems to have taken place with logic. While the Greeks and Indians paid early and detailed attention to formal logic the Chinese (as we [Needham, 1956] showed many times) exhibited a constant tendency to develop dialectical logic. The corresponding Chinese philosophy of organism paralleled Greek and Indian mechanical atomism. In these fields the 'West' was 'elementary' and China 'advanced'. And the characteristic Chinese passion for the concrete fact would have

Here, then, are some of the answers to the doubts and uncertainties of the writers who have discussed Chinese mathematics in the past, generally without the benefit of acquaintance with the original texts. Better informed observers have drawn attention to certain particular weaknesses which we can now more fully appreciate. Mikami considered that the greatest deficiency in old Chinese mathematical thought was the absence of the idea of rigorous proof, and correlated this (as do some modern Chinese scholars, such as the late Fu Ssu-Nien) with the failure of formal logic to develop in China, and with the dominance of associative (organic) thought. Cajori, evaluating, in his history of mathematical notations, the Thien Yuan algebra, says that one is impressed both by its beautiful symmetry and by its extreme limitations. After an initial burst of advance, the Sung science of algebra did not experience rapid and extended growth. The standstill which it reached after the thirteenth century he attributes to an inelastic and cramping notation. Strangely, in a people who carried algebra so far, the equational form remained implicit, and there was no indigenous development of an equality sign (=). How far the widespread use of counting-board and abacus acted as an inhibiting factor is a moot point; they certainly allowed calculations to vanish without trace, leaving no record of the intermediate stages by which the answer was reached. But it seems hard to believe that what was essentially a mechanical aid to calculation would not have been helpful if more modern mathematical methods had developed. Moreover, although so advanced in many respects (for example, the very early – c. fourth century BC – appreciation of decimal-place-value and the zero 'blank') no symbolic way of writing formulae was ever spontaneously invented by Chinese mathematicians, and until the time of the coming of the Jesuits, mathematical statements were mainly written out in characters.

But one cannot discuss the early history of algebra without deciding what is meant by the term. If we thus designate the art which allows us to solve such an equation as $a^2+bx+c = 0$ expressed in these symbols, it is a sixteenth-century development.

---

been as much inimical to the abstractness of Greek geometry as it was to the metaphysical idealism of the Buddhists. Both were flights from the practical and the empirical, the concrete and the real.

If we allow other less convenient symbols, it goes back to the third century at least; if purely geometrical solutions are allowed, it begins in the third century BC; and if we are to class as algebra any problem which would now be solved by algebraic methods, then the second millennium BC was acquainted with it. So far D. E. Smith in his excellent history of mathematics. Now algebra was dominant in Chinese mathematics as early as we can study it (about the second century BC), yet it does not fall into any of these categories. It was in fact 'rhetorical' (to use Nesselmann's term), and positional, using symbols (as generally understood) only rarely and late. In other words it brought into play an abundance of abstract monosyllabic technical ideograms indicating generalized quantities and operations. If these were not yet symbols in the mathematical sense, they were more than merely words in the ordinary sense. And then in the course of the work the counting-board with its numbers (formed by little rods placed in different positions) was laid out in such a way that certain places were occupied by specific kinds of quantities (unknowns, powers, etc.). Thus a permanent filing-system of mathematical patterns was established – quite in accord with the tendency towards organic thinking which was so overwhelmingly strong in all Chinese philosophy (see Needham, 1956, pp. 292, 336). But since the types of equations always retained their connection with concrete problems, no general theory of equations developed. However, the predilection for thinking in terms of patterns finally evolved from the counting-board a positional notation so complete (as far as it went) that it rendered unnecessary most of our fundamental symbols. Unfortunately, though the achievement was magnificent, it led to a position from which no further advance was possible.

The social background of Chinese mathematics

Turning to social factors, it is striking that throughout Chinese history the main importance of mathematics was in relation to the calendar. It would be hard to find a mathematician in the *Chhou Jen Chuan* (*Lives of the Mathematicians*, by Juan Yuan, 1799) who was not called upon to remodel the calendar of his time, or to help in such work. For reasons connected with the ancient corpus of cosmological beliefs, the establishment of the calendar was the jealously guarded prerogative of the emperor, and its acceptance on the part of tributary states signified loyalty to him. When rebellions or famines occurred, it was often concluded that something was wrong with the calendar, and the mathematicians were asked to reconstruct it. It has been thought that this preoccupation fixed them irretrievably to concrete number, and prevented the consideration of abstract ideas; but in any case the practical and empirical genius of the Chinese tended in that direction. In the calendrical field mathematics was socially orthodox and Confucian, but there are reasons for thinking that it also had unorthodox Taoist connections. In the second century Hsü Yo was certainly under Taoist influence, and this can be detected in the mysterious eleventh-century book which inspired Li Yeh; moreover, there was Hsiao Tao-Tshun's strange figure in the Sung. What one misses, however, is any contact between such personalities as Ko Hung the great alchemist and Sun Tzu the mathematician, in all probability his contemporary; doubtless such thought-connections would have been impossible anywhere before the Renaissance.[3] Lastly, a factor of great importance must be sought in the Chinese attitude to 'laws of Nature'. This has been studied in detail at the end of our second volume (Needham, 1956). Here it need only be repeated that the absence of the idea of a creator deity, and hence of a supreme law-giver, together with the firm conviction (expressed by Taoist philosophers in high poetry of Lucretian vigor) that the whole universe was an organic, self-sufficient system, led to a concept of all-embracing Order in which there was no room for Laws of Nature, and hence, no fixed regularities to which it would be profitable to apply mathematics in the mundane sphere.

If we cast a brief backward glance over the twenty centuries of autochthonous Chinese mathematics, we find that the two dynasties which stand out for mathematical achievement are the Han and the Sung. For the first century BC, the time of Lohsia

3. There was a distinctly romantic element about the Taoists, ensconced in their temples among the mountains and forests. Though they busied themselves with alchemical furnaces, they also inspired poets. But the mathematicians seem to have been very plain practical men, men in the retinues of provincial officials. Their style of writing was quite unliterary. Unlike Indian mathematical knowledge, that of China was never enshrined in verses. No doubt the Chinese mathematician also had his beautiful and intelligent Lilavati – but he kept her out of his books.

Hung and Liu Hsin, the *Chiu Chang Suan Shu* (*Nine Chapters on the Mathematical Art*) was a splendid body of knowledge. It dominated the practice of Chinese reckoning-clerks for more than a millennium. Yet in its social origins it was closely bound up with the bureaucratic government system, and devoted to the problems which the ruling officials had to solve (or persuade others to solve). Land mensuration and survey, granary dimensions, the making of dykes and canals, taxation, rates of exchange – these were the practical matters which seemed all-important. Of mathematics 'for the sake of mathematics' there was extremely little. This does not mean that Chinese calculators were not interested in truth, but it was not abstract systematized irrelevent truth after which sought the Greeks. During all this time the masses of the people remained illiterate, having no access to the manuscript books which the government commissioned, copied and distributed to the various nodes of the administrative network. Artisans, no matter how greatly gifted, remained upon the other side of an invisible wall which separated them from the scholars of literary training. It was only because Shen Kua (in the eleventh century) was so exceptional a man that he took notice of the *Mu Ching* (*Timberwork Manual*) of a great architect Yü Hao, who had probably had to dictate it to a scribe. But other artisans, some centuries before, under the significant inspiration of Taoists and Buddhists, had taken a decisive step to explode this situation – they had invented printing. This without doubt fostered the second flowering of Chinese mathematics, in the Sung, when a group of truly great mathematicians, themselves either commoners or subordinate officials, broke out into fields much wider than the traditional bureaucratic preoccupations. Intellectual curiosity could now be abundantly satisfied. But the upsurge did not last. The Confucian scholars who had practiced calligraphy on all the last copies of Tsu Chhung-Chih's *Chui Shu*[4] swept back into power in the nationalist reaction of the Ming,[5] and mathematics was again confined to the back rooms of provincial yamens. When

4. *The Art of Threading*, a mathematical book of the 5th century, long lost, but universally regarded by medieval Chinese mathematicians as of great importance. We cannot be sure of its contents.

5. Significantly, the Pa Ku essay examination system, with all its deadening influence, was introduced first in 1487.

the Jesuits entered upon the scene, there was no one even able to tell them of China's past mathematical glories.

Origins of the method of modern natural science in Europe

What was it, then, that happened at the Renaissance in Europe whereby mathematized natural science came into being? And why did this not occur in China? If it is difficult enough to find out why modern science developed in one civilization, it may be even more difficult to find out why it did not develop in another. Yet the study of an absence can throw bright light upon a presence. The problem of the fruitful union of mathematics with science is, indeed, only another way of stating the whole problem of why it was that modern science developed in Europe at all.

Pledge hits the target when he contrasts Galileo (1564 to 1642 – who must be considered the central figure in the mathematization of natural science) with Leonardo da Vinci (1452 to 1519) saying that in spite of all the latter's deep insight into nature and brilliance in experimentation, no further development followed because of his lack of mathematics. Now Leonardo was not the isolated genius that many have been led to suppose him; he was, as Zilsel, Gille and others have shown, the most outstanding of a long line of practical men in the fifteenth and sixteenth centuries – artist-engineers and architects, such as Brunelleschi (1377 to 1446); artist-metallurgists such as Cellini (1500 to 1571); gunners such as Tartaglia (1500 to 1557); surgeons such as Ambroise Paré (1510 to 1590); miners who found a voice in Agricola (1490 to 1555); shipbuilders such as those of the Arsenal at Venice which was the setting for Galileo's *Discourse* of 1638; gunpowder-millers and other chemical technologists whom Biringuccio (*d.* 1538) represents; and instrument-makers such as Robert Norman (*fl.* 1590), whose *Newe Attractive* of 1581 greatly stimulated William Gilbert's work on the magnet. All these men busied themselves with the investigation of natural phenomena, and most of them produced experimental data ready, as it were, for the magic touch of mathematical formulation. Roughly speaking, they had their Chinese counterparts:[6] such as Sung Ying-Hsing (*fl.* 1637), author of the

6. This is a very important point, for those who say, quite rightly, that China never produced a Galileo, a Vesalius or a Descartes, generally forget that China did produce men of a type similar to Agricola, Gesner and Tartaglia.

Thien Kung Khai Wu (*The Exploitation of the Works of Nature*, 1637), who may be called the Chinese Agricola; or an architect like Li Lhieh (*d.* 1110), author of *Ying Tsao Fa Shih* (*Treatise on Architectural Technique*, 1101); or the prince of pharmacists, Li Shih-Chen (1522 to 1596), author of *Pên Tshao Kang Mu* (*The Great Pharmacopoedia*, 1596); or the horticulturist Chhen Hao-Tzu (*fl.* 1688), author of the *Hua Ching* (*Horticultural Manual*, 1688); or the gunner Chiao Yü (*fl.* 1412), author of the *Huo Lung Ching* (*Firearms Manual*). Whatever field one chooses will yield parallels; thus in horological engineering (allowing for the difference of a century) a de Dondi (1318 to 1389) can be matched by a Su Sung (1020 to 1101), author of *Hsin I Hsiang Fa Yao* (*New Design for an Astronomical Clock*, 1090). But in Europe, unlike China, there was some influence at work for which this stage was not enough. Something pushed forward beyond it to make the junction between practical knowledge, empirical even when quantitatively expressed, and mathematical formulations.

Part of the story undoubtedly concerns the social changes in Europe which made the association of gentlemen with the technicians respectable. As Gabriel Harvey wrote in 1593:

> He that remembereth Humphrey Cole a mathematicall mechanician, Matthew Baker a shipwright, John Shute an architect, Robert Norman a navigator, William Bourne a gunner, John Hester a Chymist, or any like cunning and subtile empirique, is a proud man, if he contemn expert artisans, or any sensible industrious practitioners, howsoever unlectured in Schooles or unlettered in Bookes.[7]

William Gilbert's treatise on the magnet in 1600 was the first printed book composed by an academically trained scholar which was based entirely on personal manual laboratory experimentation and observation. Yet it neither employed mathematical formulations nor spoke in terms of laws of Nature. His contemporary, Francis Bacon, though the first writer in the history of mankind to realize fully the basic importance of modern scientific research for the advancement of human civilization, understood no better the enormous part which mathematics was soon to play.

Not the mathematics of the Middle Age, however. As Koyré has put it in his brilliant essay on the origins of Newtonian Cartesian science, mathematics itself had to be transformed, mathe-

7. (See Taylor, 1930).

matical entities had to be brought nearer to physics, subjected to motion, viewed not in their 'being' but in their 'becoming' or 'flux'. The calculus was the crowning achievement of this movement. In 1550 European mathematics had been hardly more advanced than the Arabic inheritance of Indian and Chinese discoveries. But there followed an astounding range of things basically new – the elaboration of a satisfactory algebraic notation at last by Vieta (1580) and Recorde (1557), the full appreciation of what decimals were capable of by Stevin (1585), the invention of logarithms by Napier (1614) and the slide rule by Gunter (1620), the establishment of coordinate and analytic geometry by Descartes (1637), the first adding machine (Pascal, 1642), and the achievement of the infinitesimal calculus by Newton (1665) and Leibniz (1684). No one has yet fully understood the inner mechanism of this development. It has often been said that whereas previously algebra and geometry had evolved separately, the former among the Indians and the Chinese and the latter among the Greeks and their successors, now the marriage of the two, the application of algebraic methods to the geometric field, was the greatest single step ever made in the progress of the exact sciences. It is important to note, however, that this geometry was not just geometry as such, but the logical deductive geometry of Greece. The Chinese had always considered geometrical problems algebraically, but that was not the same thing.

The Galilean method and its world-view

The birth of the experimental-mathematical method, which appeared in almost perfect form in Galileo, and which led to all the developments of modern science and technology, presents the history of science with one of its most important and complex questions. Though we cannot do it justice, a brief analysis here will not be out of place, for only in this way can we gain some idea of how it was exactly that mathematics and science came together at the Renaissance, and how far they had remained apart in earlier medieval (as in Chinese) society.[8] If we dissect the Galilean

8. Apart from the classical studies which will be referred to in the following pages, it is now possible to profit from the brief but valuable expositions of Dingle and Lilley. One Chinese thinker has reviewed the field – Lin Chi-Kai. He was a pupil of Abel Rey, and though he must often have had

method we find that it comprised the following phases:

1. Selection, from the phenomena under discussion, of specific aspects expressible in quantitative terms.

2. Formulation of a hypothesis involving a mathematical relationship (or its equivalent) among the quantities observed.

3. Deduction of certain consequences, from this hypothesis, which were within the range of practical verification.

4. Observation, followed by change of conditions, followed by further observation – i.e. experimentation; embodying, in so far as possible, measurement in numerical magnitudes.

5. Acceptance, or rejection, of the hypothesis framed in 2. An accepted hypothesis then served as the starting-point for fresh hypotheses and their submission to test.[9]

That the 'new, or experimental, philosophy' was characterized by the search for measurable elements in phenomena, and the application of mathematical methods to these quantitative regularities, has long been recognized. A world of quantity was substituted for the world of quality. But the advance into abstraction went further than this, for motion was considered apart from any particular moving bodies.[10] The motion of a body had no longer anything to do with its other characteristics or qualities, and could not be derived from them. This was indeed a fundamental change in outlook, for the 'uniformization' of the Cosmos was also, in a sense, its extinction and death. The geometrization of space, the substitution of homogeneous, abstract, dimensional, Euclidean space, for the concrete and differentiated place-continuum of pre-Galilean physics and astronomy, was the liquidation of what had been a morphological Cosmos.[11] In

---

Chinese parallels and problems in mind, they were rigidly excluded (unfortunately) from his thesis.

9. Here gradually, with growing confidence, entered in the element of scientific 'prediction'.

10. And Galileo did not hesitate to use concepts of the unobserved and the unobservable – such as a perfectly frictionless plane, or the motion of a body in empty infinite space.

11. Time also became continuous, undifferentiated and homogeneous, in contrast with the separate and divided times of medieval thought, both

fact, the world was no longer to be conceived of as a finite and hierarchically ordered whole, qualitatively and ontologically differentiated; but as an open, indefinite, even infinite, universe, held together only by the identity and universal applicability of simple fundamental laws. There was nowhere in the universe, for example, where the writ of the law of gravitation would not run – once the concept of gravitation had been formulated.

It is evident that the denial of 'inherent tendencies' of bodies to move towards certain places was but one aspect of a general breaking up of the organic unity of the material object. It seemed, as Dingle says, such an obvious unity, with its compact properties of shape, weight, color and movement, that only a mind of the highest originality, goaded by centuries of frustration, could take the revolutionary step of rejecting this unity, and maintaining that a wooden ball and a planet of unknown substance had more in common than the motion and the color of the same ball. And, indeed, the Galilean revolution did destroy the organic world-view which medieval Europeans had possessed, to some extent in common with the Chinese, replacing it by a world-view essentially mechanistic, and fully ripe for fortuitous concourses of atoms. The sense of loss and disorientation experienced by minds steeped in the traditional world-view was expressed by John Donne (see Manley, 1963).

And new Philosophy calls all in doubt,
The Element of Fire is quite put out;
The Sun is lost, and th' earth, and no man's wit
Can well direct him where to looke for it.
And freely men confesse that this world's spent,
When in the Planets, and the Firmament
They seeke so many new. . . .
'Tis all in Peeces, all cohaerance gone;
All just Supply, and all Relation. . . .

---

eastern and western (see Needham, 1956, p. 288). There is an echo of this in current discussions among historians concerning Chinese historiography. As van der Sprenkel has said, most Chinese writers (though by no means all) worked with a compartmentalized chronology of dynastic periods, reigns and reign-periods; discrete units of time being inhabited by particular events. The time-continuum of modern history is derived from the world-outlook of modern science.

But the dramatic irony of fate brought it about that by the time of the death of Newton himself (1727) the seeds of a new organic view of the world, destined ultimately to replace or to correct the mechanistic view, had been sown by Leibniz.[12] Perhaps some of these originated in China, but that is another argument which cannot be retraced here. (See Needham, 1956, sects. 13 et seq., 16 et seq., 18.)

That the hypothesis formed should be a mathematized hypothesis (phase 2 above) was of enormous importance. Mathematics was the largest and clearest body of connected logical thought then available. That the logic of experimentation was not absolutely bound to mathematical expression became, no doubt, apparent in the science of physiology from William Harvey and J. B. van Helmont to Claude Bernard, and in chemistry too. But it formed the model. Much discussion has centered around the origins of the mathematization of hypotheses, and the historical problem is still far from its solution. Burtt, in a well-known book, and Koyré, have emphasized the persistence of Pythagorean and Platonic influences, exemplified in the view of the universe as a mathematical design, and mediated through such men as Ficino and Novara. There was also the perennial importance of mathematics in astronomical science, and perhaps a certain stimulus from the rediscovery of Greek writers such as Archimedes. Galileo himself certainly said:

Philosophy is written in that great book which ever lies before our gaze – I mean the universe – but we cannot understand it if we do not first learn the language and grasp the symbols in which it is written. The book is written in the mathematical language, and the symbols are triangles, circles, and other geometrical figures, without the help of which it is impossible to conceive a single word of it, and without which, one wanders in vain through a dark labyrinth.[13]

Nevertheless, E. W. Strong has convincingly shown that before Galileo, and during his lifetime, mathematics was increasingly utilized by the practical technicians and artisans of whom we have already spoken. Some of these, such as Nicolò Tartaglia and

12. This is well appreciated by Koyré, who says that the evolution of nineteenth-century scientific thought, proceeding under the aegis of the field concept was an essentially anti-Newtonian development.

13. *Opera*, vol. 4, p. 171.

Simon Stevin, were among the best mathematicians of their time. Their interests in gunnery, in shipbuilding, in hydraulic engineering and building technology, invited them at all points to apply to their problems the quantitative and the mathematical. Perhaps the Galilean innovation may best be described as the marriage of craft practice with scholarly theory. The great craftsmen of the Renaissance would have found natural enough Galileo's isolation of particularly simple examples of motion in phase 3 above, and the measurement of numerical magnitudes in phase 4. In fact, as Whitehead has said, the idea of functionality had been born. One had to see how much change of a single specific condition corresponded to how much variation in the effect produced. 'Mathematics supplied the background of imaginative thought with which men of science approached the observation of Nature. Galileo produced formulae, Descartes produced formulae, Huygens produced formulae, Newton produced formulae.' Everyone began to draw curves showing the relations between natural phenomena, and to find equations to fit them.

The empirical component

In what way, then, had the instinctive experimentation of the technologists and craftsmen differed from the conscious experimental test of precise hypotheses which formed the essence of the Galilean method? The question is of great importance, for the higher artisanate (as we might perhaps call it) included as many Chinese as Europeans. Dissected in the same way as before, it might give us something like this:

1. Selection, from the phenomena under discussion, of specific aspects.

2. Observation, followed by change of conditions, followed by further observation – i.e. experimentation, embodying, in so far as possible, measurement in numerical magnitudes.

3. Formulation of a hypothesis of primitive type (e.g. involving the Aristotelian elements, the Tria Prima of the alchemists, or the Yin-Yang and Five-Element theories).

4. Continued observation and experimentation, not too strongly influenced by the concurrent hypothetical considerations.

In such empirical ways it was possible to accumulate great stores of practical knowledge, though the lack of rationale necessitated a handing down of technical skill from one generation to the next, through personal contact and training. With due regard to different times and places there was not much to choose between China and Europe regarding the heights of mastery achieved; no westerners surpassed the metallurgists of the Shang and Chou, or equalled the ceramists of the Thang and Sung. The preparation for Gilbert's definitive study of magnetism had all taken place at the other end of the Old World. And it could not be said that these technological operations were non-quantitative, for the ceramists could never have reproduced their effects in glaze and body and color without some kind of temperature control, and the discovery of magnetic declination could not have occurred if the geomancers had not been attending with some care to their azimuth degrees.

But the inhibition lay in the realm of hypothesis-making, as one may see in the relative theoretical backwardness of Leonardo. Duhem, after describing some of his achievements and inventions connected with matter in the gaseous state, points out that his ideas on air and fire, smoke and vapor, were so impregnated with medieval physics that what he did and suggested seems almost inexplicable. While sketching a hygrometer, a helicopter, or a centrifugal pump, he was capable of explaining that the moisture of a wet rag has an intrinsic tendency to move to the fire, and that its less material parts accompany the ascent of that pure element towards the empyrean, for fire has a quasi-spiritual power of carrying light things up with it. There is no need to illustrate this point more abundantly, but it is an important one, since it helps us to understand what remarkable technical achievements may be effected without adequate scientific theory. It therefore throws light on the Chinese situation, and defines the point reached by indigenous Chinese science and technology as Vincean, not Galilean. But what exactly happened in Europe remains difficult to unravel.

The speculative component

Historians have long recognized that the middle of the twelfth century was a turning-point in the history of European thought. Whether or not because of the stimulus of new contacts with the

Islamic world,[14] the twelfth and thirteenth centuries saw a vast movement away from anthropocentric symbolism towards genuine interest in objective Nature. This can be traced in every department of thought and art, from the growing naturalism of Gothic stone-carving to the rise of new realism in theology, liturgy and drama. It is not possible to overlook this naturalistic movement in tracing the roots of modern science.

The higher artisanate was not the only group which possessed part of the Galilean method before Galileo. It has long been maintained that within European scholastic philosophy there was a trend towards experimentation which, starting from Aristotle, led to Galileo through Leonardo. Aristotle distinguished[15] between knowledge of facts and knowledge of the reasons or causes of facts, but never gave any clear account of how these could be ascertained by experiment. Early in the thirteenth century philosophers at Oxford began to interest themselves in the possibilities of a deeper understanding of natural phenomena, and much attention was given to the forming of hypotheses and the manner of their testing. These ideas came later to be associated with the university at Padua, where Averroism was strong and logic was studied as a preliminary to medicine, not law or theology. Discussions there, between the fourteenth and the sixteenth centuries, led to a methodological theory which, except for the important element of mathematization, showed some similarity to the eventual practice of Galileo. The theory (called at Padua *regressus*), dissected, looks something like this:

1. Selection, from the detailed phenomena under discussion, of features which seemed to be common to all of them (*analysis*,[16] *resolutio*), complete enumeration being recognized as unnecessary because of faith in the uniformity of Nature and the representativeness of samples.

2. Induction of a specific principle by reasoning on the essential content of these features (also *resolutio*).

14. There is, at any rate, an exact contemporaneity here with such important translators as Adelard of Bath, whose chief work ended about 1142.

15. See Aristotle in references.

16. These were the earliest terms – used by the Greek geometers and Galen.

3. Deduction (*synthesis* in thought, *compositio*) of the detailed consequences of this hypothetical principle.

4. Observation of the same, and perhaps also similar, phenomena, leading to *verificatio* or *falsificatio* by experience, and in rare cases, by arranged experiment.

5. Acceptance, or rejection, of the hypothetical principle formulated in 2.

European scholastics

Thus while the practice of the higher artisanate was akin to the second or experimental part of the Galilean method, the theorizing of the scholastics foreshadowed the first or speculative part. But how widely they were aware that agreement with empirical fact was the ultimate test of hypotheses seems doubtful, nor is it clear that they always understood the importance of examining new phenomena in phase 4 which had not already been used as the origin of the hypothesis under test. Moreover, they rarely succeeded in advancing beyond the primitive style in their hypotheses. Robert Grosseteste of Lincoln (1168 to 1253) has been selected as the key figure in this natural philosophy, but the dual process of induction and deduction goes back to Galen and the Greek geometers, probably reaching Grosseteste through Arabic sources, such as the encyclopaedist Abu Yusuf Ya'qub ibn-Ishaq al-Kindi (*d.* 873) and the medical commentator 'Ali ibn Ridwan (998 to 1061). Though Grosseteste may have believed that organized experimentation beyond mere further experience should be used to verify or disprove hypotheses, it is not claimed that he himself was an experimentalist. He does seem however to have influenced the thirteenth-century group of practical scientific workers which included the Englishmen Roger Bacon (1214 to 1292) and Thomas Bradwardine (1290 to 1349) in physics, the Frenchman Petrus Peregrinus (*fl.* 1260 to 1270) in magnetism, the Pole Witelo (*c.* 1230 to 1280) in optics, and the German Theodoric of Freiburg (*d.* 1311) with his admirable theory of the rainbow.[17] It is curious that during the period in which these men

17. In which he had been anticipated by Arab physicists, but only by a very short time. Since their work was not available in Latin, the simultaneity probably arose from the use of a common source, Ibn al-Haitham.

were working, China was the scene of a scientific movement quite comparable.[18] But after the early years of the fourteenth century there was a marked regression, and verbal argument again dominated in Europe until the time of Galileo himself. Or such at any rate was the case so far as theoretical science was concerned, for the latter part of the fourteenth century and the fifteenth saw the rise of the military engineers, largely German, whose practical achievements foreshadowed those of the higher artisanate in the century before Galileo. The originator of this new phase was Konrad Kyeser (1366 to after 1405) with his *Bellifortis*, begun in 1396, but earlier figures had shown the way, especially Guido da Vigevano (1280 to after 1345) whose work on machines and war machines had been finished only a couple of decades after the death of Theodoric of Freiburg. Kyeser was followed by many other technologists who certainly owed part of their inspiration to the new techniques of gunnery and gunpowder,[19] such as Giovanni de' Fontana (*fl.* 1410 to 1420), the anonymous engineer of the Hussite Wars (*fl.* 1420 to 1433), Abraham of Memmingen (*fl.* 1422), and so on. There was thus a continuous line of experimentalists in Europe from Roger Bacon to Galileo, but after about 1310 the contribution of scholastic philosophy ceased, and for three centuries practical technology was the order of the day.[20]

18. It included the outstanding mathematicians Chhin Chiu-Shao (1240 to 1260) and Yang Hui (*fl.* 1260 to 1275), the astronomer Kuo Shou-Ching (1231 to 1316), the geographer Chu Ssu-Pên (*c.* 1270 to 1330), as well as Wang Chen. (*fl.* 1280 to 1315), encyclopaedist of agriculture and engineering, and Sung Tzhu (*fl.* 1240 to 1250), the founder of forensic medicine – a particularly experimental discipline. The period also saw the travels of men such as Marco Polo (in Peking, *c.* 1280), but it is not likely that any of them were sufficiently versed in the sciences to transmit anything more than techniques or fragments of techniques.

19. Which had been a direct Chinese transmission to Europe.

20. The late fourteenth and fifteenth centuries were not great periods of scientific or technological achievement in China, where the eleventh had been relatively much more important than in Europe. But it is not difficult to find men of worth – such as the prince Chu Hsiao (*fl.* 1382 to 1425) – who maintained a botanical garden and wrote a valuable work on plants suitable for food in emergency, or the astronomer Huangfu Chung-Ho (*fl.* 1437), who made new instruments similar to those of Kuo Shou-Ching, or Ma Huan (*fl.* 1400 to 1430), Yunnanese Muslim interpreter, geographer on the staff of the admiral Chêng Ho.

Chinese Neo-Confucians

Someone should raise the question whether the theorizing of the Neo-Confucians of the eleventh and twelfth centuries about the acquisition of natural knowledge was not as advanced in its way as that of the thirteenth-century European scholastics. The induction of a specific principle from many observations (the second part of *resolutio*) was represented in Neo-Confucian thought by the search for the underlying or intricate patterns (*Li*). Someone said to Hsü Hêng (1209 to 1281):

> If we fully apprehend [literally 'exhaust'] the patterns of the things of the world will it not be found that every thing must have a reason why it is as it is? And also a rule [of co-existence with all other things] to which it cannot but conform? Is not this just what is meant by Pattern?

Hsü Hêng agreed, saying that this brought out very well the meanings of the technical terms employed. All the spatio-temporal relations of all organisms and events in the universe were determined by the ubiquitous manifestations of *Li*. 'Wherever there is *Li*,' said Chhêng I-Chhuan (1033 to 1108) 'east is east and west is west.' The key phrase of the Neo-Confucians for the process akin to induction was taken from the *Ta Hsüeh* (*Great Learning*), a text of about 260 BC: 'The extension of knowledge consists in the investigation of things.' For them this meant a kind of sudden insight into the natures and relations of things, as if the components of a pattern were seen suddenly to 'fall into place.' The figures of Nature fitted themselves, as it were, into a meaningful array on the counting-board of the universe.

Apprehension of specific natural patterns present in a multiplicity of phenomena was reached by a process of 'relating' or 'threading together' (*kuan*), or 'inter-relating' (*kuan thung*). The image was that of holed cash threaded on a string. As one of the Chhêng brothers said, 'Whenever men hear a saying or learn of an affair, and their knowledge is still confined to this one saying or affair, it is simply because they cannot inter-relate.' Another of the formulations of the Chhêngs was the following:

> In laboring to apprehend [literally 'exhaust'] patterns fully, we are not necessitated to attempt an exhaustive and complete research into the patterns of all the myriad phenomena in the world. Nor can we attain our aim by fully apprehending only a single one of these patterns. It is

simply necessary to accumulate [literally 'pile up and tie together,' *chi lei*] a large number (of phenomena). Then [the patterns] will become visible spontaneously.

The conviction of the Chhêng brothers that concentration on one thing or on one small group of things was not the way to natural knowledge is particularly interesting in view of the failure of Chinese scholars in later times to appreciate the scientific method.[21]

Someone asked Chhêng I-Chhuan, 'Is it necessary to investigate all things, or can the innumerable patterns be known by the investigation of a single thing?' [I-Chhuan replied]: 'No, indeed, for in that case how could there be comprehensive interrelation? Even a Yen Tzu [the favorite pupil of Confucius] would not attempt to understand the patterns of all things by the investigation of only one thing. What is necessary is to investigate one thing after another day after day. Then after long accumulation of experience [the things] will reveal themselves in a state of inter-relatedness (*kuan thung*).'

Introspection was no substitute for the study of external nature.

Again someone asked Chhêng I-Chhuan, 'In observing outer things and in searching into the self, should one look back into oneself to seek what has already been seen in things?' [He replied]: 'There is no need to put the matter in that way. The external world and the self have one single great pattern in common; as soon as "that" is understood, "this" becomes clear. This is the Tao of the union of the internal and the external. The scholar should try to observe and understand all Nature – at one extreme the height of the heavens and the thickness of earth – at the other, why a single [tiny] thing is as it is.'

Somebody else said: 'In extending knowledge, what do you say to seeking [the patterns of the world] first in the "Four Beginnings" (*ssu tuan*)?'[22] [The philosopher answered]: 'To seek them in our own nature and passions is of course simple and near at home, but every [blade of] grass and every tree have their own patterns, and these must be investigated.'

A wide knowledge of the names (and properties) of birds, animals,

21. We are, of course, thinking of the idealist school of Wang Yang-Ming in the early sixteenth century; (see Needham, 1956, p. 510).

22. The reference is to Mencius, II (1), vi. 5. The feeling of commiseration is the beginning of human-heartedness, that of shame and dislike the beginning of righteousness; modesty and complaisance lead to good customs, and deciding between 'is' and 'is not' to knowledge. Cf. *Hsiao Hsüeh Kan Chu* ch. 3, p. 16a.

plants and trees, is one of the means of reaching an understanding of pattern.[23]

If this is not yet the natural science of the Renaissance, it seems no further away from it than the ideas of the medieval European scholastics.

Besides, the converse process of deduction from principles (*compositio*) seems to have its counterpart in 'extending the pattern-principle' (*thui li*) or 'enlarging the class [of things or processes with the same pattern]' (*thui lei*). Sometimes the latter phrase is used in the sense of 'inferring by analogy.' Chhêng Ming-Tao (1032 to 1085) wrote:

The myriad things all have their opposites; there is alternation of the Yin and the Yang, the good and the bad. When the Yang waxes the Yin wanes, when good increases evil is diminished. Far and wide is the spread of this pattern-principle (*Li*).

And elsewhere:

In investigating things to apprehend fully their patterns, there is no question of completely exhausting all the phenomena in the world. If the pattern is fully apprehended in one matter only, inferences can be made about other matters of the same class.

Both processes (induction followed by deduction) seem to be mentioned in these words of Chhêng I-Chhuan:

To learn [the patterns] from what is outside, and to grasp them within, may be called 'understanding' (*ming*). To grasp them within and to connect them (*chhien*) with what is outside, may be called 'integration' (*chheng*). Now integration and understanding are one.

When one comes to verification or invalidation by experiment the Neo-Confucians were no more enlightened than the Scholastics. But the idea of submission to the test of experience was always vaguely present, and the form which it took in the Chinese milieu, so dominated by ethics and sociology, was the contrast of knowledge and practice. This subject was debated throughout the

23. This was the form made famous by Sun Yat-Sen in modern China, and used, e.g. as the device of the National Peiping Academy. It is the opposite of the phrase as it made its first appearance – in one of the spurious (i.e. fourth century) chapters of the *Shu Ching* (*Historical Classic*), ch. 17B (Yüeh Ming). The speaker was Fu Yüeh, a semi-legendary minister of the Shang period.

centuries. The famous tag, 'Action is easy but knowledge difficult' (*hsing i chih nan*), was affirmed, reversed, or modified in every age. (See Needham, 1956, p. 170.) 'Knowledge,' said Wang Chhuan-Shan in the seventeenth century, 'is the beginning of practice, and practice is the completion of knowledge.' In fact the epistemological problem received answers which varied in accordance with the current trend (so far as Chinese thought allowed of it) towards metaphysical idealism or materialism. Many examples of these have been given, as in the criticism of Wang Chhung (first century) on the Mohists; while in our own time the solutions of the Chinese philosophical schools have received a Marxist critique.[24] In any case, there was no clear understanding, either among the Neo-Confucians or the Scholastics, that in the study of nature precise hypotheses must be put to the test of agreement with further ranges of empirical fact. The important point is that just as in China there were representatives of the 'higher artisanate' in abundance, counterparts of Norman and Tartaglia, so also there were medieval thinkers who corresponded to Grosseteste and the Paduans.

Perhaps it would not be fruitful to compare too closely the European Scholastics with the Neo-Confucians. At all events the former do not always pre-empt our sympathy. Two strangely contrasting statements may indicate which of the two schools was really the more scientifically minded. Thomas Aquinas (1226 to 1274) wrote that 'a little knowledge about the highest things is better than the most abundant knowledge about things low and small.' But Chhêng Ming-Tao (1032 to 1085) said of the Buddhists: 'When they strive only to "understand the high" without "studying the low", how can their understanding of the high be right?'

The social matrix in Europe and China

No question is more difficult than that of historical causation. Yet the development of modern science in Europe in the sixteenth and seventeenth centuries has either to be taken as miraculous or to be explained, even if but provisionally and tentatively. This development was not an isolated phenomenon; it occurred *pari passu* with the Renaissance, the Reformation, and the rise of

24. As in Mao Tsê-Tung's 'On Practice', and the interesting commentary of Fêng Yu-Lan upon it.

mercantile capitalism followed by industrial manufacture. It may well be that concurrent social and economic changes supervening only in Europe formed the milieu in which natural science could rise at last above the level of the higher artisanate, the semi-mathematical technicians. The reduction of all quality to quantities, the affirmation of a mathematical reality behind all appearances, the proclaiming of a space and time uniform throughout all the universe; was it not analogous to the merchant's standard of value? No goods or commodities, no jewels or monies there were, but such as could be computed and exchanged in number, quantity and measure.

Of this there are abundant traces among our mathematicians. The first literary exposition of the technique of double-entry book-keeping is contained in the best mathematical text-book available at the beginning of the sixteenth century, the *Summa de Arithmetica* (1494) of Luca Pacioli. The first application of double entry book-keeping to the problems of public finance and administration was made in the works of the engineer-mathematician Simon Stevin (1608). Even Copernicus wrote on monetary reform (in his *Monetae Cudendae Ratio* of 1552). The book of Robert Recorde, in which the equality symbol was first used (*Whetstone of Witte*, 1557) was dedicated to 'The Governors and the reste of the Companie of Venturers into Moscovia', with the wish for 'continuall increase of commodities by their travell'. Stevin's *Disme* opens with the words 'To all astronomers, surveyors, measurers of tapestry, barrels and other things, to all mintmasters and merchants, good luck!' Such examples could be indefinitely multiplied. Commerce and industry were 'in the air' as never before.

The problem of the exact relations between modern science and technology and the socio-economic circumstances of its birth constitutes, perhaps, the Great Debate of the history of science in Europe. I believe that in due course the study of parallel civilizations such as that of agrarian-bureaucratic China will throw some light on the events which took place in the West. For example, Koyré, in criticizing the socio-economic theory of the causation of post-Renaissance science,[25] urges, with Cassirer, that there was a

25. Nevertheless, he sees very clearly how congruent the mechanistic conception of tortuitous concourses of atoms was with the social ideas of

purely theoretical current stimulated by the rediscovery of Greek mathematics, and under recognizable Platonic and Pythagorean inspiration. This is doubtless part of the truth. He also urges that the supporters of the socio-economic theory insufficiently allow for what he calls the autonomous evolution of astronomy. But here is the kind of point where it is profitable to compare parallel events in China. The Chinese should have been interested in mechanics for ships, in hydrostatics for their vast canal system (like the Dutch), in ballistics for guns (after all, they had possessed gunpowder three or four centuries before Europe), and in pumps for mines. If they were not, could not the answer be sought in the fact that little or no private profit was to be gained from any of these things in Chinese society, dominated by its imperial bureaucracy? Their techniques and industries, described in the books of men akin to the writers of the 'semi-mathematical' stage mentioned above, were all essentially 'traditional', the product of many centuries of slow growth under bureaucratic oppression or at best tutelage, not the creations of enterprising merchant-venturers with big profits in sight. As for astronomy, no organization stood more in need of it than the Chinese imperial court, which by immemorial custom gave forth the calendar to be accepted by all under heaven. And, as will be seen in our third volume (Needham, 1959), Chinese astronomy was far from negligible. If an 'autonomous evolution' of astronomy was ever going to give rise to the mathematization of natural science, it is hard to see why this did not occur, or had not already occurred, in China. If the need had been sufficiently great, there would surely not have been wanting those who could have burst the bonds of the old mathematical notation, and made the discoveries which in fact were only made in Europe. But this was evidently not the force from which modern science could spring, and indigenous Chinese mathematics went down into a kind of tomb, from which the filial care of Mei Ku-Chhêng and his successors only later succeeded in resurrecting it.

Put in another way, there came no vivifying demand from the

---

capitalism, in which harmony would prevail, it was thought, if every man devoted himself to the selfish pursuit of gain. A book such as Malynes' *Lex Mercatoria* shows how conscious this analogy was in the seventeenth century.

side of natural science. Interest in Nature was not enough, controlled experimentation was not enough, empirical induction was not enough, eclipse prediction and calendar calculation were not enough – all of these the Chinese had. Apparently a mercantile culture alone was able to do what agrarian bureaucratic civilization could not – bring to fusion point the formerly separated disciplines of mathematics and nature-knowledge.

References

ARISTOTLE (1960), *Posterior Analytics and Topica*, trans. H. Tredennick and E. S. Forster, Heinemann.

MANLEY, F. (1963) (ed), 'Anatomie of the World', *The Anniversaries*, John Hopkins Press.

NEEDHAM, J. (1956), *Science and Civilisation in China*, vol. 2, Cambridge University Press.

NEEDHAM, J. (1959), *Science and Civilisation in China*, vol. 3, Cambridge University Press.

TAYLOR, F. R. G. (1930), *Today Geography, 1485–1583*, London.

WHITEHEAD, A. N. (1948), *Essays in Science and Philosophy*, London.

2 J. Ben-David and A. Zloczower

The Growth of Institutionalized Science in Germany

Excerpt from J. Ben-David and A. Zloczower, 'Universities and academic systems in modern societies', *European Journal of Sociology*, vol. 3, 1962, pp. 45–84.[1]

For about a hundred years, between the early nineteenth century and the advent of Nazism, German universities served as model academic institutions. The education of an American or British scientist was not considered complete until he had spent some time in Germany, studying with one of the renowned professors, far more of whom had won acclaim and scientific distinction than the scientists of any other country. The still prevalent conception or 'idea' of the university, as well as the definition of the professor's role, originated in Germany during the nineteenth century. It was, furthermore, in the German universities, more than anywhere else, that the main fields of scientific enquiry developed into 'disciplines' possessing specialized methodologies and systematically determined contents. Students who wanted to know what a discipline really was had to read German textbooks and those who wanted to keep abreast of scientific research had to read German journals.

The outside world, which became aware of the excellence of German achievements, connected these achievements with the internal structure and organization of the German universities. It came to be widely believed that what a university should be and how a university should be run was discovered in Germany. The discovery was – and still is – often attributed to the ideas of German philosophers from Kant to Hegel who conceived of the university as a seat of original secular learning pursued as an end in itself, and who imparted to it supreme dignity. During the nineteenth century reforms were introduced following the German example in Britain, France and the U.S., leading invariably to a rising standard

1. This excerpt is reproduced without the detailed historical footnotes provided by the authors. [Ed.]

of scientific work and a growing volume of production. This confirmed the belief that the peculiar ideas and arrangements of German universities accounted for their excellence.

We shall attempt to show that these ideas and arrangements were not the cause but rather the result of the circumstances which had historically shaped the German university; that it was not the idea of the university which explains the success of the German university system, nor the diffusion of this idea abroad which explains the impetus to science in those countries introducing organizational reforms under its impact. In order to do this we shall have to examine the circumstances which determined the strength as well as the weaknesses of the German university system.

The pioneering period of the German universities, marked by the rapid development of the different academic fields and their differentiation into systematic and specific disciplines lasted until about the end of the nineteenth century. By about 1860 the original four faculties of theology, philosophy, law and medicine, comprising just about all higher knowledge existing at the beginning of the century had been transformed beyond all recognition. A host of new disciplines had found their place within the loose frame of the faculties, none of which – with the exception of theology – seems to have been averse to incorporating new fields. Commencing with the third quarter of the century this process of expansion and differentiation slowed down. Neither the emerging social sciences nor the various fields of engineering attained proper academic status at the universities. The latter was banished to the *Technische Hochschulen*, which only over the strenuous opposition of the universities attained the right to confering the title 'doctor'. The universities not only began to offer increasing resistance to the introduction of new sciences which had mushroomed outside their walls, they also placed often insurmountable obstacles on the path of disciplines which had begun to develop organically within the established disciplines. Where previously it had been relatively easy to carve out new disciplines from the broad fields and gain recognition through the establishment of separate chairs for them, new specialities were increasingly condemned to permanent subordinate status under the pretext of being too narrow or shallow, and, therefore, '*nicht ordinierbar*'. The division of labor which arose in the *Instituten* (research laboratories usually attached to a university chair but not properly integrated within the university) raced far ahead of the increasingly out-of-date academic division of labor. The unity of teaching and research broke down when the academic scientist was forced to specialize in the *Institut* in research that threatened to isolate him from the main discipline which he had to teach if he wanted to become a full professor. The usual rule that each discipline was represented by only one professor contributed much in the previous decades to the establishment of new chairs, because the expansion of the academic staff could take place only in this manner. After the development of the institutes, however, the same rule became a veritable strangling-noose: research could be conducted only in the *Institut* but only one person, the director, could be professor.

Thus gradually a fence was drawn around the existing academic fields, excluding an increasing part of scientific and scholarly enquiry from the universities. Originally the university was meant to embrace all intellectual enquiry. It absorbed all existing disciplines, theoretical and practical. Even its philosophical founding fathers, like Fichte, Schleiermacher, and Schelling, were as much publicists as 'academic' philosophers. In the middle of the century the university became more strictly academic, but it created new disciplines and enlarged its scope, so that practically all the important scientific activity of that time originated at the university. Towards the end of the century both processes of extending the scope of the university came to a standstill.

This growing resistance to differentiation within, and to intellectual (or practical) influence from without, was accompanied by inflexibility of the organizational structure. The professorial role, and the career pattern *Privatdozent-Professor*, so well suited to the needs of research and teaching at the beginning and in the middle of the nineteenth century, when techniques and organization of research were simple, became unable to carry any more the whole burden of research and teaching. *Privatdozentur* in particular became an anomaly in fields where the most necessary research facilities were open only to assistants in the *Institut*, so that a *Privatdozent* without a position in an *Institut* had no opportunity for doing scientific work. The main career-line became, therefore, the assistantship. Yet the constitution of the university and its official structure of roles has hardly taken note of the change. Officially, even today the institutes are only appendages facilitating the professor's research. Even if in some cases this arrangement works well (depending on the personality of the professors and the nature of the discipline) it shows extreme traditionalism and ritualistic clinging to organizational forms which no longer reflect the changed functions of the university.

The explanation of this contradiction between the innovative vitality of the early years of the German university and its subsequent rigidity lies mainly in two circumstances: the fact that the German cultural area extending over the major part of Central Europe has always exceeded the limits of any German state, and the position of the university in Germany's class system. Due to the first factor there did not arise in Germany central national universities, like Paris in France, or Oxford-Cambridge in England. The university system was decentralized and competitive. Universities tried to outdo each other, or, at any rate, had to keep pace with each other. As a result innovations were introduced in Germany more easily and accepted more widely than elsewhere. The second factor, namely the position of the university in the class system, accounts for the inflexibility of its organizational structure which became manifest late in the nineteenth century. As it will be shown later, the status and the freedom of the university, seemingly so well established and secure, were as a matter of fact precarious, engendering fear of and resistance to any organizational change.

This interpretation of the developments is not in accordance with the usually accepted view which relates the rise of the German universities to the reforms introduced early in the century under the influence of the then current philosophical ideas, especially the establishment of the University of Berlin. We have to see, therefore, what was the share of ideas, and of competition and class structure, in the process. Indeed, Berlin was the first university in which the philosophical faculty (including arts and sciences) obtained a status formally equivalent, but in influence superior, to the old faculties of law, medicine and theology. There is no doubt that the granting of academic status to the new arts and sciences was a decisive step, and that philosophers had a great part in this innovation. There was a growing class of intellectuals in Germany towards the end of the eighteenth century who would not enter any more into the clergy as people like them had done before, and who interested themselves in the broad field of learning and methodical thinking which was called at that time philosophy. They were seeking social recognition and economic security, but these were unattainable for them under the existing circumstances: the bourgeoisie was relatively poor and backward, most of the aristocracy had no tradition of education, and the minority who had such interests, preferred French to German education. The only career open to young German intellectuals was a university appointment which, in the philosophical faculty, carried little prestige and did not allow real freedom of thought and speech, since universities were subject to the double control of the state and the church. Partly as a result of this control, universities were also intellectually poor institutions. They were harshly criticized, as French universities had been prior to the Revolution, and there was a tendency among enlightened circles to replace them with specialized professional schools. As a matter of fact, quite a few universities were closed down and some professional schools were established during this period.

The Napoleonic wars gave a new chance to the philosopher-intellectuals as well as to the universities. Their advocacy of German instead of French culture, unheeded before, became now the popular ideology affecting even the French-educated upper class. There was a feeling that the real strength of the nation was in the realm of spirit and culture. Indeed, after their subjugation by Napoleon, Germans had little else left to fight with but spiritual strength. This seemed all the more so because political and military defeat coincided with an unprecedented flowering of German philosophy and literature. Philosophers now became national figures, and education was given high priority. Under these circumstances the philosophical faculty was given its full university status. Since at the same time, and for the same reasons, secondary education was also reformed (through the introduction of the *Abitur*), the new faculty had plenty of students preparing for teaching in the *Gymnasium*.

Undoubtedly these reforms, which grew up in response to this constellation of circumstances, gave an important impetus to academic work, especially in philosophy and the humanities, the subjects most in fashion at that time in Germany. But only this initial impetus can be attributed to the philosophical ideas attending the birth of this new type of university. All the upsurge of the various *Fachwissenschaften*, especially of the natural sciences, occurred not as a result, but rather in spite of these ideas. The intentions of the founders and the ideologists of Berlin University would only have made it into a unique showpiece radiating light to all corners of Germany and attracting students and scholars to the somewhat provincial Prussian capital. The decisive thing, however, in the transformation of higher education and research in Germany was the fact that exactly the opposite happened. Berlin never became a unique center but rather the archaic little universities which had been hovering for decades on the verge of dissolution became transformed within a short period of time into institutions modelled on the example of Berlin. In addition a number of new universities were founded. This quite unintended success was the result of the decentralization of the German academic life. The universities, competing with each other, had to follow the successful example established in one university.

Instead of asserting, therefore, that philosophy created the new German university, we propose that the German university system provided the basis for the great development of philosophy as a systematic discipline. But contrary to the intention of the philosophers, the university system made philosophy into just one of the academic disciplines, and added to it a great many new ones. The competitive system worked according to a logic of its own. The establishment of new universities, the raising of the status of the philosophical faculties and the firm establishment of the new type of philosophy in them, created a widespread demand for philosophers in a system comprising more than twenty universities. The student of 'philosophy' in Berlin who habilitated himself could take his choice: Bonn, Greifswald, Königsberg, Göttingen, Iena were all in the market for the bright young scholar, offering professorships in the new philosophical faculties of the reorganized university system. By 1840 the philosophical faculty – fifty years previously a mere preparatory part of the theological faculty – was by far the largest in its number of teachers comprising nearly half of all the professors, *extraordinarii* and *Privatdozenten* (270 out of 633; 124 out of 253 and 142 out of 326 respectively). In addition, there was a demand for the philosophically trained person in the theological, legal and even the medical faculties which, in the latter case, prevented the development of empirical approaches for quite a while.

During the first twenty to thirty years after the reform of the universities, the general intellectual approach which hardly distinguished between philosophy, history, literature and even natural sciences was broken down into specialized disciplines: history, linguistics, philology etc. All these were closely connected with the ideological bias of German philosophy which identified culture mainly with the humanities. But the breaking down of 'philosophy' into specialized disciplines was in itself a departure from the ideological bias, and it occurred as the result of a simple mechanism; whenever the demand for professors in a certain field was saturated, there was a tendency among the more enterprising students to enter new fields regarded until then as mere sub-specialities of an established discipline, and to develop the speciality into a new discipline. Thus when the humanities were saturated around 1830–40, there occurred a shift of interest towards empirical natural sciences, and the interest in speculative philosophy at the universities abated.

This process has been traced in the development of physiology. Lectures in the subject were held already during the first decades of the nineteenth century at German universities, but work was sporadic and haphazard. In 1828 physiology as an experimental discipline was represented in only six German universities by seven lecturers. It was a side-line of anatomy which was beginning to separate from surgery, staking out the entire field of theoretical medicine as its domain. This process took place during the 30s and 40s and by the end of that period anatomy had become the main discipline of scientific medicine, the nucleus from which medicine was turned into a natural science. This new anatomy was taught at almost all German universities in connection with physiology. Competence in physiology, rather than in surgery, became a necessary qualification for attaining the chairs for anatomy. It was however not always possible to implement this requirement. Vacant chairs for anatomy could be staffed with scholars familiar with and competent in both fields, but what was one to do with the anatomists of the old school whose privilege to teach anatomy could not be revoked, who yet would not, and could not, teach the new subject? To establish separate chairs for physiology was easier than the creation of a second chair, part of whose function would be to duplicate work entrusted already to the incumbent of a recognized discipline. The forties and fifties were thus periods when specialization in physiology was encouraged and scientific activity in this sphere stimulated. Scholars who hoped for calls to chairs in anatomy were encouraged to focus their research on physiological problems, since most universities still hoped to entrust the teaching of both physiology and anatomy to a single professor, while here and there separate chairs for physiology had already established the complete independence of the new discipline. The prospect of separate chairs stimulated those scientists with special aptitude to devote themselves entirely to the new discipline. Their concentrated work (partly, no doubt, in response to the prospect of rapid advancement), soon disqualified the non-specialists from effective competition, and universities had to grant the demand of physiologists for the establishment of separate chairs. The separation of physiology from anatomy, which had been a temporary 'emergency solution' to cope with obstacles to the modification of the role of 'anatomists', thus became inevitable. This separation was implemented during the fifties and sixties of the nineteenth century. When in 1858 Johannes Müller died, and his chair was split into one for anatomy and one for physiology, this was not a pioneer innovation, but the rectification of an anachronism. Although the final separation in Giessen did not take place till 1891 the process of separation had been accomplished at almost all universities by 1870.

Between 1855 and 1874 twenty-six scientists were given their first appointment to chairs of physiology (sometimes still combined with anatomy). Ten of these were appointed between 1855 and 1859 alone. But therewith the discipline reached the limit of its expansion in the German university system (the number of chairs for physiology in German universities – excluding Austria and Switzerland – had reached fifteen by 1864, nineteen in 1873 and remained at twenty in 1880, 1890 and 1900). Between 1875 and 1894 only nine scholars received appointments to chairs in physiology, stepping into chairs vacated by their incumbents.

That aspiration to a professorship in physiology during the seventies and eighties was all but hopeless, is shown by the tenure of chairs at various universities throughout that period by first-generation physiologists, the generation which had in a cohort-like manner conquered the chairs which the university system was capable of providing. Du Bois-Reymond monopolized the chair in Berlin from 1858 to 1896; Brücke reigned in Vienna for four decades: 1849–90; Eckhard held the chair in Giessen from 1855 to 1891. Karl Ludwig, after more than fifteen years in the Josefinum in Vienna, Zürich and Marburg, managed to put in a further thirty (fruitful) years in Leipzig between 1865 and 1895; Karl Vierordt remained in Tübingen from 1855 till his death in 1884; Göttingen, Breslau, Bonn and Munich were held during 1860–1905, 1859–97, 1859–1910 and 1863–1908 by Meissner, Heidenhain, Pflüger and Voit respectively, while Ecker and Rollett kept the chairs in Freiburg 1850–87 and Graz 1863–1903 out of circulation. Large and small universities alike had not a single vacancy in physiology for decades, and no prospect of such an occurrence was in sight during the seventies and eighties. The result was that research in physiology lost momentum. A count of discoveries relevant to physiology in Germany shows that 321 such discoveries were made during the twenty years period of rapid expansion between 1855 and 1874 compared with 232 during the subsequent (and 168 during the preceding) twenty years. Scientific idealism notwithstanding, young scholars sought greener pastures. The number of *Privatdozenten* and extraordinary professors which had been twelve in 1864 declined to four in 1873, six in 1880 and rose again only in 1890 to thirteen. There were better ways of becoming a professor than through the study of physiology: hygiene for instance had only one chair in 1873 and grew to nineteen by 1900. Psychiatry grew from one chair to sixteen and ophthalmology from six to twenty-one during the same period, while pathology, which had only seven chairs in 1864 had reached eighteen by 1880. The enthusiasm for physiology cooled considerably.

This was the characteristic manner in which the German universities operated *as a system*, determining the life cycles of academic disciplines. It was the decentralized nature of the system and the competition among the individual units which brought about the rapid diffusion of innovations and not the internal structure of each unit, or the dominant philosophy of education. More than twenty first-rate full-time research positions in any one discipline was a huge market for early and mid-nineteenth century conditions, and an emerging discipline could attract considerable talent competing for those positions.

The same twenty positions, however, fell dismally short of the requirements of sustained scientific research under modern conditions. But the internal structure of the individual universities, bolstered by the idea of the university, allowed the perpetuation of this archaic arrangement in the face of changing conditions, and obstructed the growth of research roles capable of meeting the demands of modern science. The structural limitations of the German university remained latent so long as role-differentiation permitted the continued expansion of the academic profession, but once the *Institut* blocked this path toward professorial chairs, the inadequacy of the structure became manifest.

The reason why at that stage the structure of the university was not modified lies, as pointed out, in the class structure of Germany. In order to understand the way this affected the universities, we have to go back again to the origins. Prussia's rulers, even when, heeding the propaganda of intellectuals, they established the University of Berlin, were no intellectuals themselves. For them the professional training of lawyers, civil servants, doctors and teachers was the main function of higher education. By inclination they would have preferred the Napoleonic type of separate professional schools, and indeed had established such schools themselves earlier. They were converted to the idea of the university, since, as shown above, under the circumstances philosophy served the political interests of the nation and because this was also a reasonable decision from the point of view of their absolutistic principles. By granting corporate freedom to the universities they not only showed themselves as enlightened rulers, sympathetic to the intellectual mood of the time, but also vindicated the principle of legitimacy; the corporate rights of the university had been, after all, a medieval tradition destroyed by the French Revolution.

As a result, the newly founded University of Berlin, as well as the other universities following its example, have never been the institutions of which the philosophers had dreamt. The freedoms effectively granted to them were limited and the functions assigned to them were much more practical and trivial than desired.

First of all the influence of the state was always decisive, even where not visible. One of the simple ways through which state interference worked was the existence of government examinations for various professional titles. Formally, these examinations did not infringe the freedom of the universities to confer their own degrees, or establish their own courses of study. Since, however, the overwhelming majority of the students learnt for practical purposes, the curricula were greatly influenced by the wishes of the government. The influence was all the stronger, since the establishment of new chairs also depended on the government.

The curricula and in consequence the chairs and the faculties were, therefore, so constituted that the university was overwhelmingly a professional school. The freedom of the academic staff could only manifest itself within the framework established by the interest of the state. It manifested itself in the emphasis on basic subjects rather than practical training, and on theory rather than knowledge. This was the case even in the faculties of medicine and law, not to speak of the humanities and natural sciences. In these latter, the fact that the overwhelming majority of students prepared for secondary school teaching was only recognized in the usual combinations of disciplines studied, but not at all in the contents of the teaching. This aimed only at imparting the systematic knowledge of the disciplines, but took no account of teaching methodology, educational psychology etc., all of great importance to the future teacher.

Academic freedom furthermore manifested itself in the criteria used for appointments or promotions. Achievements in original research were considered – at least in principle – the most important criteria, even in such supposedly practical fields as for example clinical medicine, and the expert judgement of the academic staff, supposedly the most competent to judge people according to this criterion, was always one of the important bases of appointments (made actually by the state).

These circumstances then determined what was studied at

the university and how it was studied. The philosophical or, later, the systematically scientific aspects were emphasized in courses, the contents of which were determined largely by professional requirements. And the dual role of the professor, officially paid for teaching a subject to would-be professionals, but actually appointed for outstanding research (not necessarily central to his subject of teaching), arose as a result of similar compromise. The idea of the university, according to which both arrangements were considered as the best ways of promoting university study as well as research, was but an ideological justification of this practical compromise. Like all ideologies, it was used in defense of a constantly threatened position. Universities had to be on their guard, lest by being used openly for practical purposes – for which they were used as a matter of fact under the guises and compromises here described – they lose their precarious freedom of engaging in pure research. Hence the resistance to the dilution of the charismatic role of the professor chosen from the ranks of free *Privatdozenten*, by fully institutionalizing the new research and training roles growing up in the institutes. These latter looked 'dangerously' like mere bureaucratic-technical careers. For similar reasons the granting of academic standing to technology and new practical subjects was usually opposed. In brief, the freedom and the prestige of the German universities seemed to be safest when the university was kept isolated from the different classes and practical activities of society; it pursued, therefore, a policy aimed at the preservation of an esoteric and sacred image of itself.

Another important limitation of academic freedom was the fact that University professors were civil servants, and considered it a privilege to be part of this important corps. They were, therefore, expected to be loyal to the state, which under absolutist rule implied a great deal. As long as one genuinely agreed with the purposes of the rulers, this problem was not apparent. There was, therefore, a semblance of real freedom at the universities during the nationalist struggle and shortly thereafter, when on the one hand intellectuals often identified with Prussian politics, and on the other hand state-power was not very efficient. But after the middle of the century, when social problems and imperialism became the main political issues and the state increasingly efficient and powerful, the potential restraints on freedom became felt. Identification with the politics of the state often meant fanatical nationalism and obscurantism, a famous example of which was Treitschke, while opposition to it might have provoked interference by the state, as it happened in the case of social-democrats seeking academic appointments.

Thus, again, freedom had to be sought within these given limits. It was clear that under the prevailing conditions the introduction of actualities, whether in the form of philosophical publicism, or in the form of politically relevant social science, would not have led to detached discussion and the emergence of objective criteria, but to the flooding of universities by anti-democratic demagogues and the suppression of the limited amount of liberalism which existed in them. Thoughtful liberals, such as Max Weber for example, chose, therefore, the doctrine of *Wertfreiheit* of scientific enquiry. Declaring value judgements to be incompatible with true scientific enquiry and academic teaching seemed the most efficient way of ensuring freedom of discussion at the university. It was a morally respectable and logically justifiable principle which could be defended without recourse to the actual situation. But it was the actual situation which made this approach more or less the accepted doctrine of the university. It was the doctrine best suited to the maintenance of the delicate balance in a situation where free, non-utilitarian enquiry was supported and given high status by an absolutist state; and where free-thinking intellectuals taught students usually sharing the autocratic views of the rulers of the state, and preparing for government careers as civil servants, judges, prosecutors and teachers.

This doctrine of *Wertfreiheit*, most clearly formulated by Weber after the First World War, had been as a matter of fact an important guiding principle of academic thinking and action in the second half of the nineteenth century, i.e. as soon as the possible conflict between academic freedom and absolutism became acute. It explains the extreme caution and wariness towards intellectual influences coming from outside the universities, especially if these influences had some ideological implications.

The observed inflexibilities of the German university were,

therefore, the results of its precarious position in the German class structure. Intellectual enquiry in Germany did not thrive as part and parcel of the way of life of a 'middle class' of well-to-do people, whose position was based not on privilege but on achievement in various fields. It started thriving as a hot-house flower mainly on the whimsical support of a few members of the ruling class, and desperately attempted to establish wider roots in society. The universities created under the – from the point of view of the intellectuals – particularly favourable conditions of the struggle against Napoleon, were the only secure institutionalized framework for free intellectual activity in the country. The status and the privileges of the universities were granted to them by the military-aristocratic ruling class, and were not achieved as part of the growth of free human enterprise. It was, therefore, a precarious status based on a compromise whereby the rulers regarded the universities and their personnel as means for the training of certain types of professionals, but allowed them to do this in their own way and use their position for the pursuit of pure scholarship and science (which the rulers did not understand, but were usually willing to respect). The universities had to be, therefore, constantly on the defensive, lest by becoming suspected of subversion, they lose the élite position which ensured their freedom.

The idea of the German university evolved as a result of these conditions. It stressed the pursuit of pure science and scholarship as the principal function of the university, divided learning into disciplines with specialized methodologies, extolled *Wertfreiheit* in scholarly teaching and writing, was wary of applied subjects, as well as non-academic influence (including non-academic intellectual influence), and refused to grant institutional recognition to any teaching or research roles besides those of the *Privatdozent* and Professor. As it has been shown here, these ideas were not originally conceived as a means to an end. They were rather the description made into an ideology of the tactics actually employed by the universities in their struggle for maintaining their freedom and privileges.

It is true that the German universities had been highly successful in the development of the so-called pure scientific and scholarly fields. This success, however, was not due to any deliberate design or purpose on their part (according to the original idea there should have been a single German university devoted mainly to speculations in idealist philosophy), but to the unintended mechanism of competition which exploited rapidly all the possibilities for intellectual development open to the universities. That this development was largely limited to the basic fields was the result of the factors here described, as were the other aspects of growing inflexibility; the slowing down of the differentiation of existing disciplines into their unfolding specialities, as well as the ossification of university organization refusing to take proper notice of the transformation of scientific work.

Part Two
Structural and Cultural Features of Contemporary Pure Science

The only long standing research tradition in the sociology of science derives from Robert K. Merton's insights into the nature of its institutional structure. Merton's work, represented here by an early but highly influential contribution, treats science as a community of peers within which mutual evaluation and recognition of research contributions is the crucial element in maintaining the competitive cooperation essential to the generation of knowledge. Recognition is now universally accepted as an important source of social control within science and theoretical interest has come to centre on the nature of the criteria that influence its allocation. Here, Merton's analysis has been neither so productive nor so uncontroversial. He sees the evaluation of research as being predominantly guided by general standards of universalism and scepticism; the scientific community is unified by a common system of recognition and common standards for allocating recognition; disciplines differ essentially in their distinctive methods of generating knowledge claims. But no satisfactory demonstration of the role of universalism and scepticism as general norms in ongoing research practice has yet been produced.

In fact, the process of research in general has been very little studied by sociologists. Instead, they have concentrated on the structure of the research literature, informal communication patterns, and writings by scientists *about* science and scientific matters. They have sought to develop and test their theories of the social structure of science as far as is possible without involving the content of research. Merton's use of priority disputes to illuminate the institutional importance of originality and its effective recognition[1] is an example of the value of this approach, but in the hands of others it has sometimes resulted in an empty formalism.

The profound influence that Thomas Kuhn's work has begun to exert upon sociologists stems from the insights it offers into research as a social process. Put in sociological terms, Kuhn's thesis is that technical, esoteric models of procedure and interpretation are maintained within particular disciplines and provide the basis for both the practice and evaluation of research. The open rational mind is not the ideal instrument for recognising a scientific truth, as empiricist philosophers and sociologists would have it, rather an elaborately prepared conceptual and procedural frame of reference is essential. By pointing to the paradigmatic assumptions of scientists, and the way they define cohesive groups of researchers, Kuhn offers a sociological approach to the research process. The following paper summarises this and illustrates it with concrete examples; Kuhn's most recent work is more overtly sociological (see Kuhn, 1970 and Lakatos and Musgrave, 1970) but it is intelligible only as a derivation of the basic analysis included here.

Whatever the final assessment of Kuhn's work proves to be there can be no doubt that its influence at the present time is of great value. Sooner or later science has to be studied as an institution in action instead of as a black box with inputs and outputs.[2] Any encouragement to face the 'Verstehen' problems involved in this must be welcome. And, in focusing attention on the internal operation of scientific disciplines, Kuhn is facing sociology as a whole with profound questions about its own professional methods and ideals.

The contributions of Hagstrom and Mulkay are theoretical extensions of the work of Merton and Kuhn which also make reference to structural and cultural change in science. It cannot be said that their choice was difficult; the literature in this area is not extensive. Indeed Hagstrom's is the only

1. See, for example, Merton (1957) and for more recent work in the same vein, Merton (1968) and Cole (1970).

2. See Fisher's study (1966) of Invariant Theory as a social category, which illustrates both how rewarding and how difficult this is likely to be.

general empirical study of academic scientists yet produced by a sociologist. Fortunately it is methodologically unpretentious, as befits exploratory work, and its extensively used unstructured interview material, fascinating and suggestive as it is, makes it essential reading just as much as its theoretical ideas.

References

COLE, S. (1970), 'Professional standing and the reception of scientific discoveries', *Amer. J. Sociol.*, vol. 76, no. 2, pp. 286–306.

FISHER, C. S. (1966), 'The death of a mathematical theory: a study in the sociology of knowledge', *Archive for the History of the Exact Sciences*, no. 3, pp. 137–59.

KUHN, T. S. (1970), *The Structure of Scientific Revolutions*, 2nd edn., Chicago University Press.

LAKATOS, I., and MUSGRAVE, A. (1970), (eds.), *Criticism and the Growth of Knowledge*, Cambridge University Press.

MERTON, R. K. (1957), 'Priorities in scientific discovery', *Amer. sociol.* Rev., vol. 22, pp. 635–59. (Reprinted in B. Barber and W. Hirsch (eds.), *The Sociology of Science*, Free Press, 1962.)

MERTON, R. K. (1968), 'The Matthew effect in science', *Science*, vol. 169, pp. 56–63.

3 Robert K. Merton

The Institutional Imperatives of Science

R. K. Merton, 'Science and technology in a democratic order', *Journal of Legal and Political Sociology*, vol. 1, 1942. Republished in R. K. Merton, *Social Theory and Social Structure*, revised edition, Free Press, 1967, pp. 550–61, entitled 'Science and democratic social structure'.

Science, as any other activity involving social collaboration, is subject to shifting fortunes. Difficult as the very notion may appear to those reared in a culture which grants science a prominent if not a commanding place in the scheme of things, it is evident that science is not immune from attack, restraint and repression. Writing a little while ago, Veblen could observe that the faith of western culture in science was unbounded, unquestioned, unrivalled. The revolt from science which then appeared so improbable as to concern only the timid academician who would ponder all contingencies, however remote, has not been forced upon the attention of scientists and layman alike. Local contagions of anti-intellectualism threaten to become epidemic.

Science and society

Incipient and actual attacks upon the integrity of science have led *scientists to recognize their dependence on particular types of social structure*. Manifestos and pronouncements by associations of scientists are devoted to the relations of science and society. An institution under attack must re-examine its foundations, restate its objectives, seek out its rationale. Crisis invites self-appraisal. Now that they have been confronted with challenges to their way of life, scientists have been jarred into a state of acute self-consciousness: consciousness of self as an integral element of society with corresponding obligations and interests.[1] A tower of ivory becomes untenable when its walls are under

1. Since this was written in 1942, it is evident that the explosion at Hiroshima has jarred many more scientists into an awareness of the social consequences of their works.

assault. After a prolonged period of relative security, during which the pursuit and diffusion of knowledge had risen to a leading place if indeed not to the first rank in the scale of cultural values, scientists are compelled to vindicate the ways of science to man. Thus they have come full circle to the point of the re-emergence of science in the modern world. Three centuries ago, when the institution of science could claim little independent warrant for social support, natural philosophers were likewise led to justify science as a means to the culturally validated ends of economic utility and the glorification of God. The pursuit of science was then no self-evident value. With the unending flow of achievement, however, the instrumental was transformed into the the terminal, the means into the end. Thus fortified, the scientist came to regard himself as independent of society and to consider science as a self-validating enterprise which was in society but not of it. A frontal assault on the autonomy of science was required to convert this sanguine isolationism into realistic participation in the revolutionary conflict of cultures. The joining of the issue has led to a clarification and reaffirmation of the ethos of modern science.

Science is a deceptively inclusive word which refers to a variety of distinct though interrelated items. It is commonly used to denote a set of characteristic methods by means of which knowledge is certified; a stock of accumulated knowledge stemming from the application of these methods; a set of cultural values and mores governing the activities termed scientific or any combination of the foregoing. We are here concerned in a preliminary fashion with the cultural structure of science, that is, with one limited aspect of science as an institution. Thus, we shall consider, not the methods of science, but the mores with which they are hedged about. To be sure, methodological canons are often both technical expedients and moral compulsives, but it is solely the latter which is our concern. This is an essay in the sociology of science, not an excursion in methodology. Similarly, we shall not deal with the substantive findings of sciences (hypotheses, uniformities, laws), except as these are pertinent to standardized social sentiments toward science. This is not an adventure in polymathy.

The ethos of science is that affectively toned complex of values and norms which is held to be binding on the man of science.[2] The norms are expressed in the form of prescriptions, proscriptions, preferences and permissions. They are legitimatized in terms of institutional values. These imperatives, transmitted by precept and example and reinforced by sanctions are in varying degrees internalized by the scientist, thus fashioning his scientific conscience or, if one prefers the latter-day phrase, his superego. Although the ethos of science has not been codified,[3] it can be inferred from the moral consensus of scientists as expressed in use and wont, in countless writings on the scientific spirit and in moral indignation directed toward contraventions of the ethos.

An examination of the ethos of modern science is but a limited introduction to a larger problem: the comparative study of the institutional structure of science. Although detailed monographs assembling the needed comparative materials are few and scattered they provide some basis for the provisional assumption that 'science is afforded opportunity for development in a democratic order which is integrated with the ethos of science.' This is not to say that the pursuit of science is confined to democracies.[4] The most diverse social structures have provided some measure of support to science. We have only to remember that the Accademia del Cimento was sponsored by two Medicis; that Charles II claims historical attention for his grant of a charter to the Royal Society of London and his sponsorship of the Greenwich Observatory; that the Académie des Sciences was founded under the auspices of Louis XIV, on the advice of Colbert; that urged

2. On the concept of ethos, see Sumner (1965, pp. 36 et seq.); Speier (1938); Scheler (1933, vol. 1, pp. 225–62). Bayet (1931), in his book on the subject, soon abandons description and analysis for homily.

3. As Bayet (1931, p. 43) remarks: 'Cette morale [de la science] n'a pas eu ses théoriciens, mais elle a eu ses artisans. Elle n'a pas exprimé son idéal, mais elle l'a servi: il est impliqué dans l'existence même de la science.'

4. Tocqueville (1898, vol. 2, p. 51) went further: 'The future will prove whether these passions [for science], at once so rare and so productive, come into being and into growth as easily in the midst of democratic as in aristocratic communities. For myself, I confess that I am slow to believe it.' See another reading of the evidence: 'It is impossible to establish a simple causal relationship between democracy and science and to state that democratic society alone can furnish the soil suited for the development of science. It cannot be a mere coincidence, however, that science actually has flourished in democratic periods.' (Sigerist, 1938, p. 291).

into acquiescence by Leibniz, Frederick I endowed the Berlin Academy, and that the St Petersburg Academy of Sciences was instituted by Peter the Great (to refute the view that Russians are barbarians). But such historical facts do not imply a random association of science and social structure. There is the further question of the ratio of scientific achievement to scientific potentialities. Science develops in various social structures, to be sure, but which provide an institutional context for the fullest measure of development?

The ethos of science

The institutional goal of science is the extension of certified knowledge. The technical methods employed toward this end provide the relevant definition of knowledge: empirically confirmed and logically consistent predictions. The institutional imperatives (mores) derive from the goal and the methods. The entire structure of technical and moral norms implements the final objective. The technical norm of empirical evidence, adequate, valid and reliable, is a prerequisite for sustained true prediction; the technical norm of logical consistency, a prerequisite for systematic and valid prediction. The mores of science possess a methodologic rationale but they are binding, not only because they are procedurally efficient, but because they are believed right and good. They are moral as well as technical prescriptions.

Four sets of institutional imperatives – universalism, communism, disinterestedness, organized scepticism – comprise the ethos of modern science.

Universalism

Universalism[5] finds immediate expression in the canon that truth claims, whatever their source, are to be subjected to *preestablished impersonal criteria:* consonant with observation and with previously confirmed knowledge. The acceptance or rejection of claims entering the lists of science is not to depend on the personal

5. For a basic analysis of universalism in social relations, see Parsons (1951). For an expression of the belief that 'science is wholly independent of national boundaries and races and creeds', see the American Association for the Advancement of Science (1938) also Nature (1938).

or social attributes of their protagonist; his race, nationality, religion, class and personal qualities are as such irrelevant. Objectivity precludes particularism. The circumstance that scientifically verified formulations refer to objective sequences and correlations militates against all efforts to impose particularistic criteria of validity. The Haber process cannot be invalidated by a Nuremberg decree nor can an Anglophobe repeal the law of gravitation. The chauvinist may expunge the names of alien scientists from historical textbooks but their formulations remain indispensable to science and technology. However *echt-deutsch* or hundred-per-cent American the final increment, some aliens are accessories before the fact of every new technical advance. The imperative of universalism is rooted deep in the impersonal character of science.

However, the institution of science is but part of a larger social structure with which it is not always integrated. When the larger culture opposes universalism, the ethos of science is subjected to serious strain. Ethnocentrism is not compatible with universalism. Particularly in times of international conflict, when the dominant definition of the situation is such as to emphasize national loyalties, the man of science is subjected to the conflicting imperatives of scientific universalism and of ethnocentric particularism.[6] The

6. This stands as written in 1942. By 1948, the political leaders of Soviet Russia strengthened their emphasis on Russian nationalism and began to insist on the 'national' character of science. Thus, in an editorial, *Voprosy filosofii* (1948, p. 9): 'Only a cosmopolitan without a homeland, profoundly insensible to the actual fortunes of science, could deny with contemptuous indifference the existence of the many-hued national forms in which science lives and develops. In place of the actual history of science and the concrete paths of its development, the cosmopolitan substitutes fabricated concepts of a kind of supernational, classless science, deprived, as it were, of all the wealth of national coloration, deprived of the living brilliance and specific character of a people's creative work, and transformed into a sort of disembodied spirit ... Marxism-Leninism shatters into bits the cosmopolitan fictions concerning supra-class, non-national, "universal" science, and definitely proves that science, like all culture in modern society, is national in form and class in content.' This view confuses two distinct issues: first, the cultural context in any given nation or society may predispose scientists to focus on certain problems, to be sensitive to some and not other problems on the frontiers of science. This has long since been observed. But this is basically different from the second issue: the criteria of validity of claims to scientific knowledge are not matters of national taste and culture.

structure of the situation in which he finds himself determines the social role which is called into play. The man of science may be converted into a man of war – and act accordingly. Thus, in 1914 the manifesto of ninety-three German scientists and scholars – among them, Baeyer, Brentano, Ehrlich, Haber, Eduard Meyer, Ostwald, Planck, Schmoller and Wassermann – unloosed a polemic in which German, French and English men arrayed their political selves in the garb of scientists. Dispassionate scientists impugned 'enemy' contributions, charging nationalistic bias, log-rolling, intellectual dishonesty, incompetence and lack of creative capacity.[7] Yet this very deviation from the norm of universalism actually presupposed the legitimacy of the norm. For nationalistic bias is opprobrious only if judged in terms of the standard of universalism; within another institutional context, it is redefined as a virtue, patriotism. Thus by the very process of contemning their violation, the mores are reaffirmed.

Even under counter-pressure, scientists of all nationalities adhered to the universalistic standard in more direct terms. The international, impersonal, virtually anonymous character of science was reaffirmed.[8] (Pasteur: 'Le savant a une patrie, la science n'en a pas.') Denial of the norm was conceived as a breach of faith.

Universalism finds further expression in the demand that careers be open to talents. The rationale is provided by the

---

Sooner or later, competing claims to validity are settled by the universalistic facts of nature which are consonant with one and not with another theory. The foregoing passage is of primary interest in illustrating the tendency of ethnocentrism and acute national loyalties to penetrate the very criteria of scientific validity.

7. For an instructive collection of such documents, see Pettit and Leudet (1916). Félix Le Dantec, for example, discovers that both Ehrlich and Weismann have perpetrated typically German frauds upon the world of science. ('Le bluff de la science allemande.') Duhem (1915) concludes that the 'geometric spirit' of German science stifled the 'spirit of finesse'. Kellermann (1915) is a spirited counterpart. The conflict persisted into the post-war period; see Kherkhof (1933).

8. See the profession of faith by Professor E. Gley (in Pettit and Leudet, 1916, p. 181): '. . . il ne peut y avoir une vérité allemande, anglaise, italienne ou japonaise pas plus qu'une française. Et parler de science allemande, anglaise ou française, c'est énoncer une proposition contradictoire à l'idée même de science.' See also the affirmations of Grasset and Richet, *ibid.*

institutional goal. To restrict scientific careers on grounds other than lack of competence is to prejudice the furtherance of knowledge. Free access to scientific pursuits is a functional imperative. Expediency and morality coincide. Hence the anomaly of a Charles II invoking the mores of science to reprove the Royal Society for their would-be exclusion of John Graunt, the political arithmetician, and his instructions that 'if they found any more such tradesmen, they should be sure to admit them without further ado.'

Here again the ethos of science may not be consistent with that of the larger society. Scientists may assimilate caste-standards and close their ranks to those of inferior status, irrespective of capacity or achievement. But this provokes an unstable situation. Elaborate ideologies are called forth to obscure the incompatibility of caste-mores and the institutional goal of science. Caste-inferiors must be shown to be inherently incapable of scientific work, or, at the very least, their contributions must be systematically devaluated. 'It can be adduced from the history of science that the founders of research in physics, and the great discoverers from Galileo and Newton to the physical pioneers of our own time, were almost exclusively Aryans, predominantly of the Nordic race.' The modifying phrase, 'almost exclusively', is recognized as an insufficient basis for denying outcastes all claims to scientific achievement. Hence the ideology is rounded out by a conception of 'good' and 'bad' science: the realistic, pragmatic science of the Aryan is opposed to the dogmatic, formal science of the non-Aryan.[9] Or, grounds for exclusion are sought in the extra-scientific capacity of men of science as enemies of the state or church.[10] Thus, the exponents of a culture which abjures universalistic standards in general feel constrained to pay lip-

9. Stark (1936 and 1938). (The article from 1936 bears comparison with Duhem's contrast between 'German' and 'French' science.)

10. 'Wir haben sie ['marxistischen Leugner'] nicht entfernt als Vertreter der Wissenschaft, sondern als Parteigaenger einer politischen Lehre, die den Umsturz aller Ordnungen auf ihre Fahne geschrieben hatte. Und wir mussten hier um so entschlossener zugreifen, als ihnen die herrschende Ideologie einer wertfreien und vorassetzungslosen Wissenschaft ein willkommener Schutz fuer die Fortfeuhrung ihrer Plaene zu sein schien. Nicht wir haben uns an der Wuerde der freien Wissenschaft vergangen. ...' Rust (1936, p. 13).

service to this value in the realm of science. Universalism is deviously affirmed in theory and suppressed in practice.

However inadequately it may be put into practice, the ethos of democracy includes universalism as a dominant guiding principle. Democratization is tantamount to the progressive elimination of restraints upon the exercise and development of socially valued capacities. Impersonal criteria of accomplishment and not fixation of status characterize the democratic society. In so far as such restraints do persist, they are viewed as obstacles in the path of full democratization. Thus, in so far as laissez-faire democracy permits the accumulation of differential advantages for certain segments of the population, differentials which are not bound up with demonstrated differences in capacity, the democratic process leads to increasing regulation by political authority. Under changing conditions, new technical forms of organization must be introduced to preserve and extend equality of opportunity. The political apparatus designed to put democratic values into practice may thus vary, but universalistic standards are maintained. To the extent that a society is democratic, it provides scope for the exercise of universalistic criteria in science.

'*Communism*'

'Communism', in the non-technical and extended sense of common ownership of goods, is a second integral element of the scientific ethos. The substantive findings of science are a product of social collaboration and are assigned to the community. They constitute a common heritage in which the equity of the individual producer is severely limited. An eponymous law or theory does not enter into the exclusive possession of the discoverer and his heirs, nor do the mores bestow upon them special rights of use and disposition. Property rights in science are whittled down to a bare minimum by the rationale of the scientific ethic. The scientist's claim to 'his' intellectual 'property' is limited to that of recognition and esteem which, if the institution functions with a modicum of efficiency, is roughly commensurate with the significance of the increments brought to the common fund of knowledge. Eponymy – *e.g.*, the Copernican system, Boyle's law – is thus at once a mnemonic and a commemorative device.

Given such institutional emphasis upon recognition and esteem as the sole property right of the scientist in his discoveries, the concern with scientific priority becomes a 'normal' response. Those controversies over priority which punctuate the history of modern science are generated by the institutional accent on originality.[11] There issues a competitive cooperation. The products of competition are communized,[12] and esteem accrues to the producer. Nations take up claims to priority,[13] and fresh entries into the commonwealth of science are tagged with the names of nationals: witness the controversy raging over the rival claims of Newton and Leibniz to the differential calculus. But all this does not challenge the status of scientific knowledge as common property.

The institutional conception of science as part of the public domain is linked with the imperative for communication of

11. Newton spoke from hard-won experience when he remarked that '[natural] philosophy is such an impertinently litigious Lady, that a man had as good be engaged in lawsuits, as have to do with her.' Robert Hooke, a socially mobile individual whose rise in status rested solely on his scientific achievements, was notably 'litigious'.

12. Marked by the commercialism of the wider society though it may be, a profession such as medicine accepts scientific knowledge as common property. See Shryock (1938, p. 45) '... the medical profession ... has usually frowned upon patents taken out by medical men.... The regular profession has ... maintained this stand against private monopolies ever since the advent of patent law in the seventeenth century.' There arises an ambiguous situation in which the socialization of medical practice is rejected in circles where the socialization of knowledge goes unchallenged.

13. Now that the Russians have officially taken up a deep reverence for the Motherland, they come to insist on the importance of determining priorities in scientific discoveries. Thus: 'The slightest inattention to questions of priorities in science, the slightest neglect of them, must therefore be condemned, for it plays into the hands of our enemies, who cover their ideological aggression with cosmopolitan talk about the supposed non-existence of questions of priority in science, *i.e.* the questions regarding which peoples made what contribution to the general store of world culture.' And further: 'The Russian people has the richest history. In the course of this history it has created the richest culture, and all the other countries of the world have drawn upon it and continue to draw upon it to this day.' (*Voprosy filosofii*, 1948, pp. 10, 12). This is reminiscent of the nationalist claims made in western Europe during the nineteenth century and Nazi claims in the twentieth. (See Stark, 1936 and 1938.) Nationalist particularism does not make for detached appraisals of the course of scientific development.

findings. Secrecy is the antithesis of this norm; full and open communication its enactment.[14] The pressure for diffusion of results is reenforced by the institutional goal of advancing the boundaries of knowledge and by the incentive of recognition which is, of course, contingent upon publication. A scientist who does not communicate his important discoveries to the scientific fraternity – thus, a Henry Cavendish – becomes the target for ambivalent responses. He is esteemed for his talent and, perhaps, for his modesty. But, institutionally considered, his modesty is seriously misplaced, in view of the moral compulsive for sharing the wealth of science. Layman though he is, Aldous Huxley's comment on Cavendish is illuminating in this connection: 'Our admiration of his genius is tempered by a certain disapproval; we feel that such a man is selfish and anti-social.' The epithets are particularly instructive for they imply the violation of a definite institutional imperative. Even though it serves no ulterior motive, the suppression of scientific discovery is condemned.

The communal character of science is further reflected in the recognition by scientists of their dependence upon a cultural heritage to which they lay no differential claims. Newton's remark – 'If I have seen farther it is by standing on the shoulders of giants' – expresses at once a sense of indebtedness to the common heritage and a recognition of the essentially cooperative and cumulative quality of scientific achievement.[15] The humility

14. See Bernal (1939, pp. 150–51) who observes: 'The growth of modern science coincided with a definite rejection of the ideal of secrecy.' Bernal quotes a remarkable passage from Réaumur (*L'Art de convertir le forgé en acier*) in which the moral compulsion for publishing one's researches is explicitly related to other elements in the ethos of science. *E.g.*, '... il y eût gens qui trouvèrent étrange que j'eusse publié des secrets, qui ne devoient pas être revelés ... est-il bien sur que nos découvertes soient si fort à nous que le Public n'y ait pas droit, qu'elles ne lui appartiennent pas en quelque sorte? ... resterait il bien des circonstances, où nous soions absolument Maítres de nos découvertes? ... Nous nous devons premièrement à notre Patrie, mais nous nous devons aussi au rest du monde; ceux qui travaillent pour perfectionner les Sciences et les Arts, doivent même se regarder commes les citoyens du monde entier.'

15. It is of some interest that Newton's aphorism is a standardized phrase which had found repeated expression from at least the twelfth century. It would appear that the dependence of discovery and invention on the existing cultural base had been noted some time before the formulations of modern sociologists. See Sarton (1935) and Ockendon (1936).

of scientific genius is not simply culturally appropriate but results from the realization that scientific advance involves the collaboration of past and present generations. It was Carlyle, not Maxwell, who indulged in a mythopoeic conception of history.

The communism of the scientific ethos is incompatible with the definition of technology as 'private property' in a capitalistic economy. Current writings on the 'frustration of science' reflect this conflict. Patents proclaim exclusive rights of use and, often, nonuse. The suppression of invention denies the rationale of scientific production and diffusion, as may be seen from the court's decision in the case of *U.S. v. American Bell Telephone Co.*: 'The inventor is one who has discovered something of value. It is his absolute property. He may withhold the knowledge of it from the public. ...'[16] Responses to this conflict-situation have varied. As a defensive measure, some scientists have come to patent their work to ensure its being made available for public use. Einstein, Millikan, Compton, Langmuir have taken out patents.[17] Scientists have been urged to become promoters of new economic enterprises (Bush, 1934). Others seek to resolve the conflict by advocating socialism (Bernal, 1939, pp. 155 et seq.) These proposals – both those which demand economic returns for scientific discoveries and those which demand a change in the social system to let science get on with the job – reflect discrepancies in the conception of intellectual property.

Disinterestedness

Science, as is the case with the professions in general, includes disinterestedness as a basic institutional element. Disinterestedness is not to be equated with altruism nor interested action with egoism. Such equivalences confuse institutional and motivational levels of analysis.[18] A passion for knowledge, idle curiosity, altruistic concern with the benefit to humanity and a host of other special motives have been attributed to the scientist. The quest

16. Cited by Stern (1938). For an extended discussion, see Stern's further studies cited therein; also Hamilton (1941).

17. Hamilton (1941, p. 154); Robin (1928).

18. Parsons (1939). See Sarton (1931, p. 130 et seq.). The distinction between institutional compulsives and motives is of course a key conception of Marxist sociology.

for distinctive motives appears to have been misdirected. *It is rather a distinctive pattern of institutional control of a wide range of motives which characterizes the behavior of scientists.* For once the institution enjoins disinterested activity, it is to the interest of scientists to conform on pain of sanctions and, in so far as the norm has been internalized, on pain of psychological conflict.

The virtual absence of fraud in the annals of science, which appears exceptional when compared with the record of other spheres of activity, has at times been attributed to the personal qualities of scientists. By implication, scientists are recruited from the ranks of those who exhibit an unusual degree of moral integrity. There is, in fact, no satisfactory evidence that such is the case; a more plausible explanation may be found in certain distinctive characteristics of science itself. Involving as it does the verifiability of results, scientific research is under the exacting scrutiny of fellow-experts. Otherwise put – and doubtless the observation can be interpreted as *lèse majesté* – the activities of scientists are subject to rigorous policing, to a degree perhaps unparalleled in any other field of activity. The demand for disinterestedness has a firm basis in the public and testable character of science and this circumstance, it may be supposed, has contributed to the integrity of men of science.There is competition in the realm of science, competition which is intensified by the emphasis on priority as a criterion of achievement, and under competitive conditions there may well be generated incentives for eclipsing rivals by illicit means. But such impulses can find scant opportunity for expression in the field of scientific research. Cultism, informal cliques, prolific but trivial publications – these and other techniques may be used for self-aggrandizement. (See the account of Wilson, 1964, pp. 201 et seq.) But, in general, spurious claims appear to be negligible and ineffective. The translation of the norm of disinterestedness into practice is effectively supported by the ultimate accountability of scientists to their compeers. The dictates of socialized sentiment and of expediency largely coincide, a situation conducive to institutional stability.

In this connection, the field of science differs somewhat from that of other professions. The scientist does not stand vis-à-vis a lay clientele in the same fashion as do the physician and lawyer, for example. The possibility of exploiting the credulity, ignorance and dependence of the layman is thus considerably reduced. Fraud, chicane and irresponsible claims (quackery) are even less likely than among the 'service' professions. To the extent that the scientist-layman relation does become paramount, there develop incentives for evading the mores of science. The abuse of expert authority and the creation of pseudo-sciences are called into play when the structure of control exercised by qualified compeers is rendered ineffectual. (See Brady, 1937, ch. 2 and Gardner, 1953.)

It is probable that the reputability of science and its lofty ethical status in the estimate of the layman is in no small measure due to technological achievements.[19] Every new technology bears witness to the integrity of the scientist. Science realizes its claims. However, its authority can be and is appropriated for interested purposes, precisely because the laity is often in no position to distinguish spurious from genuine claims to such authority. The presumably scientific pronouncements of totalitarian spokesmen on race or economy or history are for the uninstructed laity of the same order as newspaper reports of an expanding universe or wave mechanics. In both instances, they cannot be checked by the man-in-the-street and in both instances, they may run counter to common sense. If anything, the myths will seem more plausible and are certainly more comprehensible to the general public than accredited scientific theories, since they are closer to common-sense experience and to cultural bias. Partly as a result of scientific achievements, therefore, the population at large becomes susceptible to new mysticisms expressed in apparently scientific terms. The borrowed authority of science bestows prestige on the unscientific doctrine.

Organized scepticism

As we have seen in the preceding chapter [not included here], organized scepticism is variously interrelated with the other elements of the scientific ethos. It is both a methodologic and an

19. Bacon (1905, p. 4) set forth one of the early and most succinct statements of this popular pragmatism: 'What is most useful in practice is most correct in theory.'

institutional mandate. The suspension of judgement until 'the facts are at hand' and the detached scrutiny of beliefs in terms of empirical and logical criteria have periodically involved science in conflict with other institutions. Science which asks questions of fact, including potentialities, concerning every aspect of nature and society may come into conflict with other attitudes towards these same data which have been crystallized and often ritualized by other institutions. The scientific investigator does not preserve the cleavage between the sacred and the profane, between that which requires uncritical respect and that which can be objectively analyzed. ('*Ein Professor ist ein Mensch der anderer Meinung ist.*')

This appears to be the source of revolts against the so-called intrusion of science into other spheres. Such resistance on the part of organized religion has become less significant as compared with that of economic and political groups. The opposition may exist quite apart from the introduction of specific scientific discoveries which appear to invalidate particular dogmas of church, economy or state. It is rather a diffuse, frequently vague, apprehension that scepticism threatens the current distribution of power. Conflict becomes accentuated whenever science extends its research to new areas toward which there are institutionalized attitudes or whenever other institutions extend their area of control. In modern totalitarian society, anti-rationalism and the centralization of institutional control both serve to limit the scope provided for scientific activity.

References

American Association for the Advancement of Science (1938), Resolution of the Council, *Science*, vol. 87, p. 10.
BACON, F. (1905), *Novum Organum*, (Book 2, p. 4), English translation by R. Ellis and J. Spedding, London.
BAYET, A. (1931), *La Morale de la Science*, Paris.
BERNAL, J. D. (1939), *The Social Function of Science*, Routledge.
BRADY, R. A. (1937), *The Spirit and Structure of German Fascism*. New York.
BUSH, V. (1934), 'Trends in engineering research', *Sigma Xi Quarterly* vol. 22, p. 49.
DUHEM, P. (1915), *La Science Allemande*, Paris.
GARDNER, M. (1953), *In the Name of Science*, Putnam's.
HAMILTON, W. (1941), 'Patents and free enterprise', *Temporary National Economics Committee Monograph, no. 31.*

KELLERMAN, H. (1915), *Der Krieg der Geister*, Weimar.
KHERKHOF, K. (1933), *Der Krieg gegen die Deutsche Wissenschaft*, Halle.
NATURE (1938), 'The advancement of science and society: proposed world association', *Nature*, vol. 141, p. 169.
OCKENDEN, R. E. (1936), *Isis*, no. 25, pp. 451–2.
PARSONS, T. (1939), 'The professions and social structure', *Social Forces*, vol. 17, pp 458–9.
PARSONS, T. (1951), *The Social System*, Free Press.
PETTIT, G., and LEUDET, M. (1916), *Les Allemands et la Science*, Paris.
ROBIN, J. (1928), *L'Oeuvre Scientifique, sa Protection – Juridique*, Paris.
RUST, B. (1936), *Das Nationalsozialistiche Deutschland und die Wissenschaft*, Hamburg.
SARTON, G. (1931), *The History of Science and the New Humanism*, Peter Smith, New York.
SARTON, G. (1935), *Isis*, no. 24, pp. 107–9.
SCHELER, M. (1933), *Schriften ans dem Nachlass*, Berlin.
SIGERIST, H. E. (1938), 'Science and democracy', *Science and Society*, vol. 2, p. 291.
SHRYOCK, R. H. (1938), 'Freedom and interference in medicine', *The Annals*, vol. 200, p. 45.
SPEIER, H. (1938), 'The social determination of ideas', *Social Research*, vol. 5, pp. 196 et seq.
STARK, J. (1936), 'Philipp Lenard als deutcher Naturforscher', *Nationalsozialistiche Monatshefte*, no. 7, pp. 106–112.
STARK, J. (1938), 'The pragmatic and dogmatic spirit in physics' Nature, vol. 141, p. 770–72.
STERN, B. J. (1938), 'Restraints upon the utilization of inventions', *The Annals*, vol. 200, p. 21.
SUMNER, W. G. (1965), *Folkways*, New English Library.
TOCQUEVILLE, A. de (1898), *Democracy in America*, New York. (Translation by Lawrence published by Fontana 1968).
VOPROSY FILOSOFII (1948), 'Against the bourgeois ideology of cosmopolitanism', as trans. in the *Current Digest of the Soviet Press*, 1949, vol. 1, no. 1.
WILSON, L. (1964), *The Academic Man*, Octagon Press.

4 Thomas S. Kuhn

Scientific Paradigms*

T. S. Kuhn, 'The function of dogma in scientific research', in
A. C. Crombie (ed.), *Scientific Change*, Heinemann, 1963, pp. 347–69.

At some point in his or her career every member of this Symposium has, I feel sure, been exposed to the image of the scientist as the uncommitted searcher after truth. He is the explorer of nature – the man who rejects prejudice at the threshold of his laboratory, who collects and examines the bare and objective facts, and whose allegiance is to such facts and to them alone. These are the characteristics which make the testimony of scientists so valuable when advertising proprietary products in the United States. Even for an international audience, they should require no further elaboration. To be scientific is, among other things, to be objective and openminded.

Probably none of us believes that in practice the real-life scientist quite succeeds in fulfilling this ideal. Personal acquaintance, the novels of Sir Charles Snow, or a cursory reading of the history of science provides too much counter-evidence. Though the scientific enterprise may be open-minded, whatever

*The ideas developed in this paper have been abstracted, in a drastically condensed form, from the first third of my monograph (Kuhn, 1962). Some of them were also partially developed in an earlier essay (Kuhn, 1959a).

On this whole subject see also Cohen (1952) and Barber (1961). I am indebted to Mr Barber for an advance copy of that helpful paper. Above all, those concerned with the importance of quasi-dogmatic commitments as a requisite for productive scientific research should see the works of Polanyi, particularly (1958) and (1951). The discussion which follows this paper will indicate that Mr Polanyi and I differ somewhat about what scientists are committed to, but that should not disguise the very great extent of our agreement about the issues discussed explicitly below.

this application of that phrase may mean, the individual scientist is very often not. Whether his work is predominantly theoretical or experimental, he usually seems to know, before his research project is even well under way, all but the most intimate details of the result which that project will achieve. If the result is quickly forthcoming, well and good. If not, he will struggle with his apparatus and with his equations until, if at all possible, they yield results which conform to the sort of pattern which he has foreseen from the start. Nor is it only through his own research that the scientist displays his firm convictions about the phenomena which nature can yield and about the ways in which these may be fitted to theory. Often the same convictions show even more clearly in his response to the work produced by others. From Galileo's reception of Kepler's research to Nägeli's reception of Mendel's, from Dalton's rejection of Gay Lussac's results to Kelvin's rejection of Maxwell's, unexpected novelties of fact and theory have characteristically been resisted and have often been rejected by many of the most creative members of the professional scientific community. The historian, at least, scarcely needs Planck to remind him that: 'A new scientific truth is not usually presented in a way that convinces its opponents . . .; rather they gradually die off, and a rising generation is familiarized with the truth from the start.' (Planck, 1948, p. 22.)

Familiar facts like these – and they could easily be multiplied – do not seem to bespeak an enterprise whose practitioners are notably open-minded. Can they at all be reconciled with our usual image of productive scientific research? If such a reconciliation has not seemed to present fundamental problems in the past, that is probably because resistance and preconception have usually been viewed as extraneous to science. They are, we have often been told, no more than the product of inevitable *human* limitations; a proper scientific method has no place for them; and that method is powerful enough so that no mere human idiosyncrasy can impede its success for very long. On this view, examples of a scientific *parti pris* are reduced to the status of anecdotes, and it is that evaluation of their significance that this essay aims to challenge. Verisimilitude, alone, suggests that such a challenge is required. Preconception and resistance seem the rule rather than the exception in mature scientific development.

Furthermore, under normal circumstances they characterize the very best and most creative research as well as the more routine. Nor can there be much question where they come from. Rather than being characteristics of the aberrant individual, they are community characteristics with deep roots in the procedures through which scientists are trained for work in their profession. Strongly held convictions that are prior to research often seem to be a precondition for success in the sciences.

Obviously I am already ahead of my story, but in getting there I have perhaps indicated its principal theme. Though preconception and resistance to innovation could very easily choke off scientific progress, their omnipresence is nonetheless symptomatic of characteristics upon which the continuing vitality of research depends. Those characteristics I shall collectively call the dogmatism of mature science, and in the pages to come I shall try to make the following points about them. Scientific education inculcates what the scientific community had previously with difficulty gained – a deep commitment to a particular way of viewing the world and of practising science in it. That commitment can be, and from time to time is, replaced by another, but it cannot be merely given up. And, while it continues to characterize the community of professional practitioners, it proves in two respects fundamental to productive research. By defining for the individual scientist both the problems available for pursuit and the nature of acceptable solutions to them, the commitment is actually constitutive of research. Normally the scientist is a puzzle-solver like the chess player, and the commitment induced by education is what provides him with the rules of the game being played in his time. It its absence he would not be a physicist, chemist, or whatever he has been trained to be.

In addition, commitment has a second and largely incompatible research role. Its very strength and the unanimity with which the professional group subscribes to it provides the individual scientist with an immensely sensitive detector of the trouble spots from which significant innovations of fact and theory are almost inevitably educed. In the sciences most discoveries of unexpected fact and all fundamental innovations of theory are responses to a prior breakdown in the rules of the previously established game. Therefore, though a quasi-dogmatic commitment is, on the one hand, a source of resistance and controversy, it is also instrumental in making the sciences the most consistently revolutionary of all human activities. One need make neither resistance nor dogma a virtue to recognize that no mature science could exist without them.

Before examining further the nature and effects of scientific dogma, consider the pattern of education through which it is transmitted from one generation of practitioners to the next. Scientists are not, of course, the only professional community that acquires from education a set of standards, tools, and techniques which they later deploy in their own creative work. Yet even a cursory inspection of scientific pedagogy suggests that it is far more likely to induce professional rigidity than education in other fields, excepting, perhaps, systematic theology. Admittedly the following epitome is biased toward the American pattern, which I know best. The contrasts at which it aims must, however, be visible, if muted, in European and British education as well.

Perhaps the most striking feature of scientific education is that, to an extent quite unknown in other creative fields, it is conducted through textbooks, works written especially for students. Until he is ready, or very nearly ready, to begin his own dissertation, the student of chemistry, physics, astronomy, geology, or biology is seldom either asked to attempt trial research projects or exposed to the immediate products of research done by others – to, that is, the professional communications that scientists write for their peers. Collections of 'source readings' play a negligible role in *scientific* education. Nor is the science student encouraged to read the historical classics of his field – works in which he might encounter other ways of regarding the questions discussed in his text, but in which he would also meet problems, concepts, and standards of solution that his future profession had long-since discarded and replaced.[1] Whitehead somewhere caught this quite special feature of the sciences when he wrote, 'A science that hesitates to forget its founders is lost.'

1. The individual sciences display some variation in these respects. Students in the newer and also in the less theoretical sciences – e.g. parts of biology, geology, and medical science – are more likely to encounter both contemporary and historical source materials than those in, say, astronomy, mathematics, or physics.

An almost exclusive reliance on textbooks is not all that distinguishes scientific education. Students in other fields are, after all, also exposed to such books, though seldom beyond the second year of college and even in those early years not exclusively. But in the sciences different textbooks display different subject matters rather than, as in the humanities and many social sciences, exemplifying different approaches to a single problem field. Even books that compete for adoption in a single science course differ mainly in level and pedagogic detail, not in substance or conceptual structure. One can scarcely imagine a physicist's or chemist's saying that he had been forced to begin the education of his third-year class almost from first principles because its previous exposure to the field had been through books that consistently violated his conception of the discipline. Remarks of that sort are not by any means unprecedented in several of the social sciences. Apparently scientists agree about what it is that every student of the field must know. That is why, in the design of a pre-professional curriculum, they can use textbooks instead of eclectic samples of research.

Nor is the characteristic technique of textbook presentation altogether the same in the sciences as elsewhere. Except in the occasional introductions that students seldom read, science texts make little attempt to describe the *sorts* of problems that the professional may be asked to solve or to discuss the *variety* of techniques that experience has made available for their solution. Instead, these books exhibit, from the very start, concrete problem-solutions that the profession has come to accept as paradigms, and they then ask the student, either with a pencil and paper or in the laboratory, to solve for himself problems closely modelled in method and substance upon those through which the text has led him. Only in elementary language instruction or in training a musical instrumentalist is so large or essential a use made of 'finger exercises'. And those are just the fields in which the object of instruction is to produce with maximum rapidity strong 'mental sets' or *Einstellungen*. In the sciences, I suggest, the effect of these techniques is much the same. Though scientific development is particularly productive of consequential novelties, scientific education remains a relatively dogmatic initiation into a pre-established problem-solving tradition that the student is neither invited nor equipped to evaluate.

The pattern of systematic textbook education just described existed in no place and in no science (except perhaps elementary mathematics) until the early nineteenth century. But before that date a number of the more developed sciences clearly displayed the special characteristics indicated above, and in a few cases had done so for a very long time. Where there were no textbooks there had often been universally received paradigms for the practice of individual sciences. These were scientific achievements reported in books that all the practitioners of a given field knew intimately and admired, achievements upon which they modelled their own research and which provided them with a measure of their own accomplishment. Aristotle's *Physica*, Ptolemy's *Almagest*, Newton's *Principia* and *Opticks*, Franklin's *Electricity*, Lavoisier's *Chemistry*, and Lyell's *Geology* – these works and many others all served for a time implicitly to define the legitimate problems and methods of a research field for succeeding generations of practitioners. In their day each of these books, together with others modelled closely upon them, did for its field much of what textbooks now do for these same fields and for others besides.

All of the works named above are, of course, classics of science. As such their role may be thought to resemble that of the main classics in other creative fields, for example the works of a Shakespeare, a Rembrandt, or an Adam Smith. But by calling these works, or the achievements which lie behind them, paradigms rather than classics, I mean to suggest that there is something else special about them, something which sets them apart both from some other classics of science and from all the classics of other creative fields.

Part of this 'something else' is what I shall call the exclusiveness of paradigms. At any time the practitioners of a given speciality may recognize numerous classics, some of them – like the works of Ptolemy and Copernicus or Newton and Descartes – quite incompatible one with the other. But that same group, if it has a paradigm at all, can have only one. Unlike the community of artists – which can draw simultaneous inspiration from the works of, say, Rembrandt *and* Cézanne and which therefore studies both – the community of astronomers had no alternative to choosing *between* the competing models of scientific activity supplied by Copernicus and Ptolemy. Furthermore, having made their choice, astronomers could thereafter neglect the work which they had rejected. Since the sixteenth century there have been only two full editions of the *Almagest*, both produced in the nineteenth century and directed exclusively to scholars. In the mature sciences there is no apparent function for the equivalent of an art museum or a library of classics. Scientists know when books, and even journals, are out of date. Though they do not then destroy them, they do, as any historian of science can testify, transfer them from the active departmental library to desuetude in the general university depository. Up-to-date works have taken their place, and they are all that the further progress of science requires.

This characteristic of paradigms is closely related to another, and one that has a particular relevance to my selection of the term. In receiving a paradigm the scientific community commits itself, consciously or not, to the view that the fundamental problems there resolved have, in fact, been solved once and for all. That is what Lagrange meant when he said of Newton: 'There is but one universe, and it can happen to but one man in the world's history to be the interpreter of its laws.'[2] The example of either Aristotle or Einstein proves Lagrange wrong, but that does not make the fact of his commitment less consequential to scientific development. Believing that what Newton had done need not be done again, Lagrange was not tempted to fundamental reinterpretations of nature. Instead, he could take up where the men who shared his Newtonian paradigm had left off, striving both for neater formulations of that paradigm and for an articulation that would bring it into closer and closer agreement with observations of nature. That sort of work is undertaken only by those who feel that the model they have chosen is entirely secure. There is nothing quite like it in the arts, and the parallels in the social sciences are at best partial. Paradigms determine a developmental pattern for the mature sciences that is unlike the one familiar in other fields.

That difference could be illustrated by comparing the development of a paradigm-based science with that of, say, philosophy or

2. Quoted in this form by Mason (1956, p. 254). The original, which is identical in spirit but not in words, seems to derive from Delambre's contemporary éloge (1816, p. 46).

literature. But the same effect can be achieved more economically by contrasting the early developmental pattern of almost any science with the pattern characteristic of the same field in its maturity. I cannot here avoid putting the point too starkly, but what I have in mind is this. Excepting in those fields which, like biochemistry, originated in the combination of existing specialities, paradigms are a relatively late acquisition in the course of scientific development. During its early years a science proceeds without them, or at least without any so unequivocal and so binding as those named illustratively above. Physical optics before Newton or the study of heat before Black and Lavoisier exemplifies the pre-paradigm developmental pattern that I shall immediately examine in the history of electricity. While it continues, until, that is, a first paradigm is reached, the development of a science resembles that of the arts and of most social sciences more closely than it resembles the pattern which astronomy, say, had already acquired in Antiquity and which all the natural sciences make familiar today.

To catch the difference between pre- and post-paradigm scientific development consider a single example. In the early eighteenth century, as in the seventeenth and earlier, there were almost as many views about the nature of electricity as there were important electrical experimenters, men like Hauksbee, Gray, Desaguliers, Du Fay, Nollet, Watson, and Franklin. All their numerous concepts of electricity had something in common – they were partially derived from experiment and observation and partially from one or another version of the mechanico-corpuscular philosophy that guided all scientific research of the day. Yet these common elements gave their work no more than a family resemblance. We are forced to recognize the existence of several competing schools and sub-schools, each deriving strength from its relation to a particular version (Cartesian or Newtonian) of the corpuscular metaphysics, and each emphasizing the particular cluster of electrical phenomena which its own theory could do most to explain. Other observations were dealt with by *ad hoc* elaborations or remained as outstanding problems for further research.[3]

3. Much documentation for this account of electrical development can be retrieved from Roller and Roller (1954) and from Cohen (1956). For analytic detail I am, however, very much indebted to a still [1963] unpublished

One early group of electricians followed seventeenth-century practice, and thus took attraction and frictional generation as the fundamental electrical phenomena. They tended to treat repulsion as a secondary effect (in the seventeenth century it had been attributed to some sort of mechanical rebounding) and also to postpone for as long as possible both discussion and systematic research on Gray's newly discovered effect, electrical conduction. Another closely related group regarded repulsion as the fundamental effect, while still another took attraction and repulsion together to be equally elementary manifestations of electricity. Each of these groups modified its theory and research accordingly, but they then had as much difficulty as the first in accounting for any but the simplest conduction effects. Those effects provided the starting point for still a third group, one which tended to speak of electricity as a 'fluid' that ran through conductors rather than as an 'effluvium' that emanated from non-conductors. This group, in its turn, had difficulty reconciling its theory with a number of attractive and repulsive effects.[4]

At various times all these schools made significant contributions to the body of concepts, phenomena, and techniques from which Franklin drew the first paradigm for electrical science. Any definition of the scientist that excludes the members of these schools will exclude their modern successors as well. Yet anyone surveying the development of electricity before Franklin may well conclude that, though the field's practitioners were scientists, the immediate result of their activity was something less than science. Because the body of belief he could take for granted was very small, each electrical experimenter felt forced to begin by building his field anew from its foundations. In doing so his choice of

---

paper by my student, John L. Heilbron, who has also assisted in the preparation of the three notes that follow.

4. This division into schools is still somewhat too simplistic. After 1720 the basic division is between the French school (Du Fay, Nollet, etc.) who base their theories on attraction-repulsion effects and the English school (Desaguliers, Watson, etc.) who concentrate on conduction effects. Each group had immense difficulty in explaining the phenomena that the other took to be basic. (See, for example, Needham's report of Lemonier's investigations (1746).) Within each of these groups, and particularly the English, one can trace further subdivision depending upon whether attraction or repulsion is considered the more fundamental electrical effect.

supporting observation and experiment was relatively free, for the set of standard methods and phenomena that every electrician must employ and explain was extraordinarily small. As a result, throughout the first half of the century, electrical investigations tended to circle back over the same ground again and again. New effects were repeatedly discovered, but many of them were rapidly lost again. Among those lost were many effects due to what we should now describe as inductive charging and also Du Fay's famous discovery of the two sorts of electrification. Franklin and Kinnersley were surprised when, some fifteen years later, the latter discovered that a charged ball which was repelled by rubbed glass would be attracted by rubbed sealing-wax or amber.[5] In the absence of a well-articulated and widely received theory (a desideratum which no science possesses from its very beginning and which few if any of the social sciences have achieved today), the situation could hardly have been otherwise. During the first half of the eighteenth century there was no way for electricians to distinguish consistently between electrical and non-electrical effects, between laboratory accidents and essential novelties, or between striking demonstration and experiments which revealed the essential nature of electricity.

This is the state of affairs which Franklin changed.[6] His theory

5. Du Fay's discovery that there are two sorts of electricity and that these are mutually attractive but self-repulsive is reported and documented in great experimental detail in the fourth of his famous memoirs on electricity (1735). These memoirs were well known and widely cited, but Desaguliers (1741) seems to be the only electrician who, for almost two decades, even mentions that some charged bodies will attract each other. For Franklin's and Kinnersley's 'surprise' see Cohen (1941, pp. 250–5). Note also that, though Kinnersley had *produced* the effect, neither he nor Franklin seems ever to have *recognized* that two resinously charged bodies would repel each other, a phenomenon directly contrary to Franklin's theory.

6. The change is not, of course, due to Franklin alone nor did it occur overnight. Other electricians, most notably William Watson, anticipated parts of Franklin's theory. More important, it was only after essential modifications, due principally to Aepinus, that Franklin's theory gained the general currency requisite for a paradigm. And even then there continued to be two formulations of the theory: the Franklin–Aepinus one-fluid form and a two-fluid form due principally to Symmer. Electricians soon reached the conclusion that no electrical test could possibly discriminate between the two theories. Until the discovery of the battery, when the choice between

explained so many – though not all – of the electrical effects recognized by the various earlier schools that within a generation all electricians had been converted to some view very like it. Though it did not resolve quite all disagreements, Franklin's theory was electricity's first paradigm, and its existence gives a new tone and flavour to the electrical researches of the last decades of the eighteenth century. The end of inter-school debate ended the constant reiteration of fundamentals; confidence that they were on the right track encouraged electricians to undertake more precise, esoteric, and consuming sorts of work. Freed from concern with any and all electrical phenomena, the newly united group could pursue selected phenomena in far more detail, designing much special equipment for the task and employing it more stubbornly and systematically than electricians had ever done before. In the hands of a Cavendish, a Coulomb, or a Volta the collection of electrical facts and the articulation of electrical theory were, for the first time, highly directed activities. As a result the efficiency and effectiveness of electrical research increased immensely, providing evidence for a societal version of Francis Bacon's acute methodological dictum: 'Truth emerges more readily from error than from confusion.'

Obviously I exaggerate both the speed and the completeness with which the transition to a paradigm occurs. But that does not make the phenomenon itself less real. The maturation of electricity as a science is not coextensive with the entire development of the field. Writers on electricity during the first four decades of the eighteenth century possessed far more information about electrical phenomena than had their sixteenth- and seventeenth-century predecessors. During the half-century after 1745 very few new sorts of electrical phenomena were added to their lists. Nevertheless, in important respects the electrical writings of the last two decades of the century seemed further removed from those of Gray, Du Fay, and even Franklin than are the writings of these early eighteenth-century electricians from those of their predecessors a hundred years before. Some time between 1740 and 1780 electricians, as a group, gained what astronomers had achieved in Antiquity, students of motion in the Middle Ages,

a one-fluid and two-fluid theory began to make an occasional difference in the design and analysis of experiments, the two were equivalent.

of physical optics in the late seventeenth century, and of historical geology in the early nineteenth. They had, that is, achieved a paradigm, possession of which enabled them to take the foundation of their field for granted and to push on to more concrete and recondite problems.[7] Except with the advantage of hindsight, it is hard to find another criterion that so clearly proclaims a field of science.

These remarks should begin to clarify what I take a paradigm to be. It is, in the first place, a fundamental scientific achievement and one which includes both a theory and some exemplary applications to the results of experiment and observation. More important, it is an open-ended achievement, one which leaves all sorts of research still to be done. And, finally, it is an accepted achievement in the sense that it is received by a group whose members no longer try to rival it or to create alternates for it. Instead, they attempt to extend and exploit it in a variety of ways to which I shall shortly turn. That discussion of the work that paradigms leave to be done will make both their role and the reasons for their special efficacy clearer still. But first there is one rather different point to be made about them. Though the reception of a paradigm seems historically prerequisite to the most effective sorts of scientific research, the paradigms which enhance research effectiveness need not be and usually are not permanent. On the contrary, the developmental pattern of mature science is usually from paradigm to paradigm. It differs from the pattern characteristic of the early or pre-paradigm period not by the total elimination of debate over fundamentals, but by the drastic restriction of such debate to occasional periods of paradigm change.

Ptolemy's *Almagest* was not, for example, any less a paradigm because the research tradition that descended from it has ultimately to be replaced by an incompatible one derived from the work of Copernicus and Kepler. Nor was Newton's *Opticks* less a paradigm for eighteenth-century students of light because it was later replaced by the ether-wave theory of Young and Fresnel, a paradigm

7. Note that this first electrical paradigm was fully effective only until 1800, when the discovery of the battery and the multiplication of electrochemical effects initiated a revolution in electrical theory. Until a new paradigm emerged from that revolution, the literature of electricity, particularly in England, reverted in many respects to the tone characteristic of the first half of the eighteenth century.

which in its turn gave way to the electro-magnetic displacement theory that descends from Maxwell. Undoubtedly the research work that any given paradigm permits results in lasting contributions to the body of scientific knowledge and technique, but paradigms themselves are very often swept aside and replaced by others that are quite incompatible with them. We can have no recourse to notions like the 'truth' or 'validity' of paradigms in our attempt to understand the special efficacy of the research which their reception permits.

On the contrary, the historian can often recognize that in declaring an older paradigm out of date or in rejecting the approach of some one of the pre-paradigm schools a scientific community has rejected the embryo of an important scientific perception to which it would later be forced to return. But it is very far from clear that the profession delayed scientific development by doing so. Would quantum mechanics have been born sooner if nineteenth-century scientists had been more willing to admit that Newton's corpuscular view of light might still have something significant to teach them about nature? I think not, although in the arts, the humanities, and many social sciences that less doctrinaire view is very often adopted toward classic achievements of the past. Or would astronomy and dynamics have advanced more rapidly if scientists had recognized that Ptolemy and Copernicus had chosen equally legitimate means to describe the earth's position? That view was, in fact, suggested during the seventeenth century, and it has since been confirmed by relativity theory. But in the interim it was firmly rejected together with Ptolemaic astronomy, emerging again only in the very late nineteenth century when, for the first time, it had concrete relevance to unsolved problems generated by the continuing practice of non-relativistic physics. One could argue, as indeed by implication I shall, that close eighteenth- and nineteenth-century attention either to the work of Ptolemy or to the relativistic views of Descartes, Huygens, and Leibniz would have delayed rather than accelerated the revolution in physics with which the twentieth century began. Advance from paradigm to paradigm rather than through the continuing competition between recognized classics may be a functional as well as a factual characteristic of mature scientific development.

Much that has been said so far is intended to indicate that – except during occasional extraordinary periods to be discussed in the last section of this paper – the practitioners of a mature scientific specialty are deeply committed to some one paradigm-based way of regarding and investigating nature. Their paradigm tells them about the sorts of entities with which the universe is populated and about the way the members of that population behave; in addition, it informs them of the questions that may legitimately be asked about nature and of the techniques that can properly be used in the search for answers to them. In fact, a paradigm tells scientists so much that the questions it leaves for research seldom have great intrinsic interest to those outside the profession. Though educated men as a group may be fascinated to hear about the spectrum of fundamental particles or about the processes of molecular replication, their interest is usually quickly exhausted by an account of the beliefs that already underlie research on these problems. The outcome of the individual research project is indifferent to them, and their interest is unlikely to awaken again until, as with parity nonconservation, research unexpectedly leads to paradigm-change and to a consequent alteration in the beliefs which guide research. That, no doubt, is why both historians and popularizers have devoted so much of their attention to the revolutionary episodes which result in change of paradigm and have so largely neglected the sort of work that even the greatest scientists necessarily do most of the time.

My point will become clearer if I now ask what it is that the existence of a paradigm leaves for the scientific community to do. The answer – as obvious as the related existence of resistance to innovation and as often brushed under the carpet – is that scientists, given a paradigm, strive with all their might and skill to bring it into closer and closer agreement with nature. Much of their effort, particularly in the early stages of a paradigm's development, is directed to articulating the paradigm, rendering it more precise in areas where the original formulation has inevitably been vague. For example, knowing that electricity was a fluid whose individual particles act upon one another at a distance, electricians after Franklin could attempt to determine the quantitative law of force between particles of electricity. Others could seek the mutual interdependence of spark length, electroscope deflection, quantity of electricity, and conductor-configuration. These were the sorts of problems upon which Coulomb, Cavendish, and Volta worked in the last decades of the eighteenth century, and they have many parallels in the development of every other mature science. Contemporary attempts to determine the quantum mechanical forces governing the interactions of nucleons fall precisely in this same category, paradigm-articulation.

That sort of problem is not the only challenge which a paradigm sets for the community that embraces it. There are always many areas in which a paradigm is assumed to work but to which it has not, in fact, yet been applied. Matching the paradigm to nature in these areas often engages much of the best scientific talent in any generation. The eighteenth-century attempts to develop a Newtonian theory of vibrating strings provide one significant example, and the current work on a quantum mechanical theory of solids provides another. In addition, there is always much fascinating work to be done in improving the match between a paradigm and nature in an area where at least limited agreement has already been demonstrated. Theoretical work on problems like these is illustrated by eighteenth-century research on the perturbations that cause planets to deviate from their Keplerian orbits as well as by the elaborate twentieth-century theory of the spectra of complex atoms and molecules. And accompanying all these problems and still others besides is a recurring series of instrumental hurdles. Special apparatus had to be invented and built to permit Coulomb's determination of the electrical force law. New sorts of telescopes were required for the observations that, when completed, demanded an improved Newtonian perturbation theory. The design and construction of more flexible and more powerful accelerators is a continuing desideratum in the attempt to articulate more powerful theories of nuclear forces. These are the sorts of work on which almost all scientists spend almost all of their time.[8]

Probably this epitome of normal scientific research requires no elaboration in this place, but there are two points that must now be made about it. First, all of the problems mentioned above were paradigm-dependent, often in several ways. Some – for example the derivation of perturbation terms in Newtonian

8. The discussion in this paragraph and the next is considerably elaborated in my paper (Kuhn, 1961).

planetary theory – could not even have been stated in the absence of an appropriate paradigm. With the transition from Newtonian to relativity theory a few of them became different problems and not all of these have yet been solved. Other problems – for example the attempt to determine a law of electric forces – could be and were at least vaguely stated before the emergence of the paradigm with which they were ultimately solved. But in that older form they proved intractable. The men who described electrical attractions and repulsions in terms of effluvia attempted to measure the resulting forces by placing a charged disc at a measured distance beneath one pan of a balance. Under those circumstances no consistent or interpretable results were obtained. The prerequisite for success proved to be a paradigm that reduced electrical action to a gravity-like action between point particles at a distance. After Franklin electricians thought of electrical action in those terms; both Coulomb and Cavendish designed their apparatus accordingly. Finally, in both these cases and in all the others as well a commitment to the paradigm was needed simply to provide adequate motivation. Who would design and build elaborate special-purpose apparatus, or who would spend months trying to solve a particular differential equation, without a quite firm guarantee that his effort, if successful, would yield the anticipated fruit?

This reference to the anticipated outcome of a research project points to the second striking characteristic of what I am now calling normal, or paradigm-based, research. The scientist engaged in it does not at all fit the prevalent image of the scientist as explorer or as inventor of brand new theories which permit striking and unexpected predictions. On the contrary, in all the problems discussed above everything but the detail of the outcome was known in advance. No scientist who accepted Franklin's paradigm could doubt that there was a law of attraction between small particles of electricity, and they could reasonably suppose that it would take a simple algebraic form. Some of them had even guessed that it would prove to be an inverse square law. Nor did Newtonian astronomers and physicists doubt that Newton's law of motion and of gravitation could ultimately be made to yield the observed motions of the moon and planets even though, for over a century, the complexity of the requisite mathematics prevented good agreement's being uniformly obtained. In all these problems, as in most others that scientists undertake, the challenge is not to uncover the unknown but to obtain the known. Their fascination lies not in what success may be expected to disclose but in the difficulty of obtaining success at all. Rather than resembling exploration, normal research seems like the effort to assemble a Chinese cube whose finished outline is known from the start.

Those are the characteristics of normal research that I had in mind, when, at the start of this essay, I described the man engaged in it as a puzzle-solver, like the chess player. The paradigm he has acquired through prior training provides him with the rules of the game, describes the pieces with which it must be played, and indicates the nature of the required outcome. His task is to manipulate those pieces within the rules in such a way that the required outcome is produced. If he fails, as most scientists do in at least their first attacks upon any given problem, that failure speaks only to his lack of skill. It cannot call into question the rules which his paradigm has supplied, for without those rules there would have been no puzzle with which to wrestle in the first place. No wonder, then, that the problems (or puzzles) which the practitioner of a mature science normally undertakes presuppose a deep commitment to a paradigm. And how fortunate it is that that commitment is not lightly given up. Experience shows that, in almost all cases, the reiterated efforts, either of the individual or of the professional group, do at last succeed in producing within the paradigm a solution to even the most stubborn problems. That is one of the ways in which science advances. Under those circumstances can we be surprised that scientists resist paradigm-change? What they are defending is, after all, neither more nor less than the basis of their professional way of life.

By now one principal advantage of what I began by calling scientific dogmatism should be apparent. As a glance at any Baconian natural history or a survey of the pre-paradigm development of any science will show, nature is vastly too complex to be explored even approximately at random. Something must tell the scientist where to look and what to look for, and that something, though it may not last beyond his generation, is the paradigm with which his education as a scientist has supplied him. Given that paradigm and the requisite confidence in it, the scientist largely ceases to be an explorer at all, or at least to be an explorer of the unknown. Instead, he struggles to articulate and concretize the known, designing much special-purpose apparatus and many special-purpose adaptations of theory for that task. From those puzzles of design and adaptation he gets his pleasure. Unless he is extraordinarily lucky, it is upon his success with them that his reputation will depend. Inevitably the enterprise which engages him is characterized, at any one time, by drastically restricted vision. But within the region upon which vision is focused the continuing attempt to match paradigms to nature results in a knowledge and understanding of esoteric detail that could not have been achieved in any other way. From Copernicus and the problem of precession to Einstein and the photo-electric effect, the progress of science has again and again depended upon just such esoterica. One great virtue of commitment to paradigms is that it frees scientists to engage themselves with tiny puzzles.

Nevertheless, this image of scientific research as puzzle-solving or paradigm-matching must be, at the very least, thoroughly incomplete. Though the scientist may not be an explorer, scientists do again and again discover new and unexpected sorts of phenomena. Or again, though the scientist does not normally strive to invent new sorts of basic theories, such theories have repeatedly emerged from the continuing practice of research. But neither of these types of innovation would arise if the enterprise I have been calling normal science were always successful. In fact, the man engaged in puzzle-solving very often resists substantive novelty, and he does so for good reason. To him it is a change in the rules of the game and any change of rules is intrinsically subversive. That subversive element is, of course, most apparent in major theoretical innovations like those associated with the names of Copernicus, Lavoisier, or Einstein. But the discovery of an unanticipated phenomenon can have the same destructive effects although usually on a smaller group and for a far shorter time. Once he had performed his first follow-up experiments, Röntgen's glowing screen demonstrated that previously standard cathode ray equipment was behaving in ways for which no one had made allowance. There was an unanticipated variable to be controlled; earlier researches, already on their way to becoming paradigms, would require re-evaluation; old puzzles would have to be solved again under a somewhat different set of rules. Even so readily assimilable a discovery as that of X-rays can violate a paradigm that has previously guided research. It follows that, if the normal puzzle-solving activity were altogether successful, the development of science could lead to no fundamental innovations at all.

But of course normal science is not always successful, and in recognizing that fact we encounter what I take to be the second great advantage of paradigm-based research. Unlike many of the early electricians, the practitioner of a mature science knows with considerable precision what sort of result he should gain from his research. As a consequence he is in a particularly favourable position to recognize when a research problem has gone astray. Perhaps, like Galvani or Röntgen, he encounters an effect that he knows ought not to occur. Or perhaps, like Copernicus, Planck, or Einstein, he concludes that the reiterated failures of his predecessors in matching a paradigm to nature is presumptive evidence of the need to change the rules under which a match is to be sought. Or perhaps, like Franklin or Lavoisier, he decides after repeated attempts that no existing theory can be articulated to account for some newly discovered effect. In all of these ways and in others besides the practice of normal puzzle-solving science can and inevitably does lead to the isolation and recognition of anomaly. That recognition proves, I think, prerequisite for almost all discoveries of new sorts of phenomena and for all fundamental innovations in scientific theory. After a first paradigm has been achieved, a breakdown in the rules of the pre-established game is the usual prelude to significant scientific innovation.

Examine the case of discoveries first. Many of them, like Coulomb's law or a new element to fill an empty spot in the periodic table, present no problem. They were not 'new sorts of phenomena' but discoveries anticipated through a paradigm and achieved by expert puzzle-solvers: that sort of discovery is a natural product of what I have been calling normal science. But not all discoveries are of that sort: many could not have been anticipated by any extrapolation from the known; in a sense they had to be made 'by accident'. On the other hand the accident through which they emerged could not ordinarily have occurred to a man just looking around. In the mature sciences discovery demands much special equipment, both conceptual and instrumental, and that special equipment has invariably been developed and deployed for the pursuit of the puzzles of normal research. Discovery results when that equipment fails to function as it should. Furthermore, since some sort of at least temporary failure occurs during almost every research project, discovery results only when the failure is particularly stubborn or striking and only when it seems to raise questions about accepted beliefs and procedures. Established paradigms are thus often doubly prerequisite to discoveries. Without them the project that goes astray would not have been undertaken. And even when the project has gone astray, as most do for a while, the paradigm can help to determine whether the failure is worth pursuing. The usual and proper response to a failure in puzzle-solving is to blame one's talents or one's tools and to turn next to another problem. If he is not to waste time, the scientist must be able to discriminate essential anomaly from mere failure.

That pattern – discovery through an anomaly that calls established techniques and beliefs in doubt – has been repeated again and again in the course of scientific development. Newton discovered the composition of white light when he was unable to reconcile measured dispersion with that predicted by Snell's recently discovered law of refraction. (See Kuhn, 1958, pp. 27–45.) The electric battery was discovered when existing detectors of static charges failed to behave as Franklin's paradigm said they should. (See Galvani, 1954, pp. 27–9). The planet Neptune was discovered through an effort to account for recognized anomalies in the orbit of Uranus. (See Armitage, 1950, pp. 111–15.) The element chlorine and the compound carbon monoxide emerged during attempts to reconcile Lavoisier's new chemistry with laboratory observations.[9] The so-called noble gases were the products of a long series of investigations initiated by a small but persistent anomaly in the measured density of atmospheric nitrogen. (See Ramsay, 1896, chs. 4, 5.) The electron was posited to explain some anomalous properties of electrical conduction through gases, and its spin was suggested to account for other sorts of anomalies observed in atomic spectra. (See Thomson,

9. For chlorine see Meyer (1891, pp. 224–7). For carbon monoxide see Kopp (1845, 294–6).

1937, pp. 325–71; Chalmers, 1949, pp. 187–217; Richtmeyer, Kennard and Lauritsen, 1955, p. 212.) Both the neutron and the neutrino provide other examples, and the list could be extended almost indefinitely. (See Richtmeyer, Kennard and Lauritsen, 1955, pp. 466–70; and R. D. Rusk, 1958, pp. 328–30.) In the mature sciences unexpected novelties are discovered principally after something has gone wrong.

If, however, anomaly is significant in preparing the way for new discoveries, it plays a still larger role in the invention of new theories. Contrary to a prevalent, though by no means universal, belief, new theories are not invented to account for observations that have not previously been ordered by theory at all. Rather, at almost all times in the development of any advanced science, all the facts whose relevance is admitted seem either to fit existing theory well or to be in the process of conforming. Making them conform better provides many of the standard problems of normal science. And almost always committed scientists succeed in solving them. But they do not always succeed, and, when they fail repeatedly and in increasing numbers, then their sector of the scientific community encounters what I am elsewhere calling 'crisis'. Recognizing that something is fundamentally wrong with the theory upon which their work is based, scientists will attempt more fundamental articulations of theory than those which were admissible before. (Characteristically, at times of crisis, one encounters numerous different versions of the paradigm theory.[10]) Simultaneously they will often begin more nearly random experimentation within the area of difficulty hoping to discover some effect that will suggest a way to set the situation right.

10. One classic example (see Kuhn, 1957, pp. 133–40), is the proliferation of geocentric astronomical systems in the years before Copernicus's heliocentric reform. Another (see Partington and McKie, 1937, 1938 and 1939) is the multiplicity of 'phlogiston theories' produced in response to the general recognition that weight is always gained on combustion and to the experimental discovery of many new gases after 1760. The same proliferation of versions of accepted theories occurred in mechanics and electromagnetism in the two decades preceding Einstein's special relativity theory. (See Whittaker, 1951–53, vol. 1, ch. 12, and vol. 2, ch. 2. I concur in the widespread judgment that this is a very biased account of the genesis of relativity theory, but it contains just the detail necessary to make the point here at issue.)

Only under circumstances like these, I suggest, is a fundamental innovation in scientific theory both invented and accepted.

The state of Ptolemaic astronomy was, for example, a recognized scandal before Copernicus proposed a basic change in astronomical theory, and the preface in which Copernicus described his reasons for innovation provides a classic description of the crisis state. (See Kuhn, 1957, pp. 133–40.) Galileo's contributions to the study of motion took their point of departure from recognized difficulties with medieval theory, and Newton reconciled Galileo's mechanics with Copernicanism.[11] Lavoisier's new chemistry was a product of the anomalies created jointly by the proliferation of new gases and the first systematic studies of weight relations.[12] The wave theory of light was developed amid growing concern about anomalies in the relation of diffraction and polarization effects to Newton's corpuscular theory. (See Whittaker, 1951–3, vol. 2, pp. 94–109; Whewell, 1847, vol. 2, pp. 213–7 and Kuhn, 1961, p. 181 n.) Thermodynamics, which later came to seem a superstructure for existing sciences, was established only at the price of rejecting the previously paradigmatic caloric theory.[13] Quantum mechanics was born from a variety of difficulties surrounding black-body radiation, specific heat, and the photo-electric effect. (See Richtmeyer et al., pp. 89–94, 124–32, and 409–14; Holton, 1953, pp. 528–45.) Again the list could be extended, but the point should already be clear. New theories arise from work conducted under old ones, and they do so only when something is observed to have gone wrong. Their prelude is widely recognized anomaly, and that recognition can come only to a group that knows very well what it would mean to have things go right.

Because limitations of space and time force me to stop at this point, my case for dogmatism must remain schematic. I shall not

11. For Galileo see Koyré (1939); for Newton see Kuhn (1957, pp. 228–60 and 289–91).

12. For the proliferation of gases see Partington (1948, ch. 6); for the role of weight relations see Guerlac (1959).

13. For a general account of the beginnings of thermodynamics (including much relevant bibliography) see Kuhn, 1959b. For the special problems presented to caloric theorists by energy conservation see the Carnot papers, there cited in n. 2, and also Thompson (1910, ch. 6).

here even attempt to deal with the fine-structure that scientific development exhibits at all times. But there is another more positive qualification of my thesis, and it requires one closing comment. Though successful research demands a deep commitment to the *status quo*, innovation remains at the heart of the enterprise. Scientists are *trained* to operate as puzzle-solvers from established rules, but they are also *taught* to regard themselves as explorers and inventors who know no rules except those dictated by nature itself. The result is an acquired tension, partly within the individual and partly within the community, between professional skills on the one hand and professional ideology on the other. Almost certainly that tension and the ability to sustain it are important to science's success. In so far as I have dealt exclusively with the dependence of research upon tradition, my discussion is inevitably one-sided. On this whole subject there is a great deal more to be said.

But to be one-sided is not necessarily to be wrong, and it may be an essential preliminary to a more penetrating examination of the requisites for successful scientific life. Almost no one, perhaps no one at all, needs to be told that the vitality of science depends upon the continuation of occasional tradition-shattering innovations. But the apparently contrary dependence of research upon a deep commitment to established tools and beliefs receives the very minimum of attention. I urge that it be given more. Until that is done, some of the most striking characteristics of scientific education and development will remain extraordinarily difficult to understand.

References

ARMITAGE, A. (1950), *A Century of Astronomy*, Sampson Low.

BARBER, B. (1961), 'Resistance by scientists to scientific discovery', Science, no. 84, pp. 596–602.

CHALMERS, T. W. (1949), *Historic Researches, Chapters in the History of Physical and Chemical Discovery*, Morgan Bros. London.

COHEN, I. B. (1941) (ed.), *Benjamin Franklin's Experiments: A New Edition of Franklin's Experiments and Observations on Electricity*, Cambridge, Mass.

COHEN, I. B. (1952), 'Orthodoxy and scientific progress', *Proceedings of the American Philosophical Society*, vol. 96, pp. 505–12.

Cohen, I. B. (1956), Franklin and Newton: *An Inquiry into Speculative Newtonian Experimental Science and Franklin's Work in Electricity as an Example Thereof*, Harvard University Press.

Delambre, J. H. B. (1816), (p. 353) eloge, *Mémoires de . . . l'Institut . . ., année 1812*, pt. 2, p. 46.

Desaguliers, J. T. (1741), (p. 356), 'Some charged bodies will attract each other', *Philosophical Transactions*, no. 42, pp. 140–3.

Du Fay, (1735), 'De l'attraction et repulsion des corps electriques', *Mémoires de . . . l'Académie . . . de l'année 1733*, Paris, pp. 457–760.

Galvani, L. (1954), *Commentary on the Effects of Electricity on Muscular Motion*, trans. M. G. Foley (notes and intro. I. B. Cohen), Nowalk, Conn.

Guerlac, H. (1959), 'The origin of Lavoisier's work on combustion', *Archives internationales d'histoire des sciences*, no. 12, pp. 113–35.

Holton, G. (1953), *Introduction to Concepts and Theories in Physical Science*, Addison-Wesley.

Kopp, H. (1945), *Geschitche der Chimie*, Braunschweig.

Koyré, A. (1939), *Etudes Galinéenes* (3 vols.), Paris.

Kuhn, T. S. (1957), *The Copernican Revolution: Planetary Astronomy in the Development of Western Thought*, Harvard University Press.

Kuhn, T. S. (1958), 'Newton's optical papers', in I. B. Cohen (ed.), *Isaac Newton's Papers & Letters on Natural Philosophy*, Cambridge, Mass.

Kuhn, T. S. (1959a), 'The essential tension: tradition and innovation in scientific research', in C. W. Taylor (ed.) *The Third (1959) University of Utah Conference on the Identification of Creative Scientific Tallent*, Salt Lake City.

Kuhn, T. S. (1959b), 'Energy conservation as an example of simultaneous discovery', in M. Clagett (ed.), *Critical Problems in the History of Science*, pp. 321–56, Madison.

Kuhn, T. S. (1961), 'The function of measurement in modern physical science', *Isis*, no. 52, pp. 161–93.

Kuhn, T. S. (1962), *The Structure of Scientific Revolutions*, University of Chicago Press.

Mason, S. F. (1956), *Main Currents of Scientific Thought*, New York. (Rev. edn 1966 as *History of the Sciences*, Collier Macmillan.)

Meyer, E. von (1891), *A History of Chemistry from the Earliest Times to the Present Day*, trans. G. McGowan, London.

Needham, H. J. T. (1746), Report of Lemonier's investigations, *Philosophical Transactions*, no. 94, p. 247.

Partington, J. R. (1948), *A Short History of Chemistry*, London, 2nd edn (prob. Macmillan).

Partington, J. R., and McKie, D. (1937, 1938, 1939), 'Historical studies of the phlogiston theory', *Annals of Science*, vols. 2, 3, 4.

Planck, M. (1948), *Wissenschaftliche Selbistbiographie*, Leipzig.

POLANYI, M. (1951), *The Logic of Liberty*, Routledge & Kegan Paul.
POLANYI, M. (1958), *Personal Knowledge*, University of Chicago Press.
RAMSAY, W. (1896), *The Gases of the Atmosphere: The History of their Discovery*, Macmillan.
RICHTMEYER, F. K., et al. (1953), *Introduction to Concerts and Theories in Physical Science*, Cambridge, Mass.
RICHTMEYER, F. K., KENNARD, E. H., LAURITSEN, T. (1955), *Introduction to Modern Physics*, McGraw-Hill, 5th edn.
ROLLER, D., and ROLLER, D. H. D. (1954), *The Development of the Concept of Electric Charge: Electricity from the Greeks to Coulomb*, Harvard Case Histories in Experimental Science, no. 8; Harvard University Press.
RUSK, R. D. (1958), *Introduction to Atomic and Nuclear Physics*, Appleton-Century Crofts.
THOMPSON, S. P. (1910), *The Life of William Thomson, Baron Kelvin of Largs* (2 vols.), Macmillan.
THOMSON, J. J. (1937), *Recollections and Reflections*, Bell.
WHEWELL, W. (1847), *History of the Inductive Sciences*, Parker, rev. edn.
WHITTAKER, E. T. (1951–53), *History of the Theories of Aether and Electricity*, Nelson, 2nd edn.

5 W. O. Hagstrom

Gift-Giving as an Organizing Principle in Science

Excerpt from W. O. Hagstrom, *The Scientific Community*, Basic Books, 1965, pp. 12–22.

Manuscripts submitted to scientific periodicals are often called 'contributions,' and they are, in fact, gifts. Authors do not usually receive royalties or other payments, and their institutions may even be required to aid in the financial support of the periodical.[1] On the other hand, manuscripts for which the scientific authors do receive financial payments, such as textbooks and popularizations, are, if not despised, certainly held in much lower esteem than articles containing original research results.

Gift-giving by scientists is thus similar to one of the most common modes of allocating resources to science, for this often takes the form of gifts from wealthy individuals or organizations. This has been true from the time of Cosimo de Medici to today, the time of the Rockefeller and Ford foundations. The gift status of moneys spent by industrial firms and governments on research is ambiguous; usually money seems to be spent with specific goals in mind, but the vast sums spent on space programs, particle accelerators, radiotelescopes, and so forth often seem like a potlatch by the community of nations. Neil Smelser (1959) has suggested that the gift mode of exchange is typical not only of science but of all institutions concerned with the maintenance and transmission of common values, such as the family, religion, and communities.

In general, the acceptance of a gift by an individual or a community implies a recognition of the status of the donor and the existence of certain kinds of reciprocal rights. (See Gouldner, 1960

1. When this is so, editorial decisions to publish are kept independent of the possibility of payment. Thus, in 1962, only 78 per cent of the pages published in the *Journal of Mathematical Physics* were paid for by the authors' institutions. (See Barton, 1963.)

and Mauss, 1954, pp. 40 et seq., 73, et passim.) These reciprocal rights may be to a return gift of the same kind and value, as in many primitive economic systems, or to certain appropriate sentiments of gratitude and deference. In science, the acceptance by scientific journals of contributed manuscripts establishes the donor's status as a scientist – indeed, status as a scientist can be achieved *only* by such gift-giving – and it assures him of prestige within the scientific community. The remainder of this chapter concerns the nature and forms of this allocation of prestige.

The organization of science consists of an exchange of social recognition for information. But, as in all gift-giving, the expectation of return gifts (of recognition) cannot be publicly acknowledged as the motive for making the gift. A gift is supposed to be given, not in the expectation of a return, but as an expression of the sentiment of the donor towards the recipient. Thus, in the kula expeditions of the Melanesians:

> The ceremony of transfer is done with solemnity. The object given is disdained or suspect; it is not accepted until it is thrown on the ground. The donor affects an exaggerated modesty. Solemnly bearing his gift, accompanied by the blowing of a conch-shell, he apologizes for bringing only his leavings and throws the objects at his partner's feet. ... Pains are taken to show one's freedom and autonomy as well as one's magnanimity, yet all the time one is actuated by the mechanisms of obligation which are resident in the gifts themselves. (Mauss, 1954, p. 21, reporting the work of Malinowski.)

Gift-giving is capable of cynical manipulation; if this is publicly expressed, however, the exchange of gifts ceases, perhaps to be succeeded by contractual exchange. Consequently, scientists usually deny that they are strongly motivated by a desire for recognition, or that this desire influences their research decisions. A biochemist gave a typical reply when asked whether scientists compete for recognition:

> Most scientists have sincere interests in the advancement of science, more than in their own recognition. I don't think honor is the greatest ambition of professionals generally; rather, they want to solve a problem. Professionals are a relatively idealistic group. You might find extremists of all sorts, but by and large their real interest is in their work and in advancing knowledge. They are motivated by curiosity. These

people think that rewards will take care of themselves; they are fatalistic in that respect. . . . There is, after all, a kind of achievement in just getting to be a professor [i.e., in attaining the status of a scholar-scientist in the larger community], which is likely to be satisfying.

Some of my informants allowed themselves to be pressed into admitting that recognition was a source of gratification. For example, a theoretical physicist responded, when told that another scientist was not disturbed at all when he was anticipated and thereby prevented from publishing:

I think I would admit to not having such pure interests. I must admit to a desire for recognition. I suppose it doesn't make much difference whether one wants to glorify himself or have others glorify him.

It is only the exceptional scientist, however, who sees the desire for recognition as a prime motivating force for himself and his colleagues. A mathematician, for example, said:

A field in mathematics may become popular if the more popular mathematicians, the big shots, become interested in it. Then it grows rapidly. Junior mathematicians want recognition from big shots and, consequently, work in areas prized by them.

This man was something of a social isolate, capable of taking a detached view of the system.[2]

Nevertheless, the public disavowal of the expectation of recognition in return for scientific contributions should no more be taken to mean that the expectation is absent than the magnanimous front of the kula trader can be taken to mean that he does not expect a return gift. In both instances, this is made clearest when the expected response is not forthcoming. In primitive societies, failure to present return gifts often means warfare. (See Mauss, 1954, p. 3 et passim). In science, the failure to recognize discovery may give rise, if not to warfare, at least to strong antagonisms

2. In a series of interviews with twenty eminent American biologists, Anne Roe was given the same impression about the suppression of the wish for recognition: '. . . the concentration is on the work primarily as an end in itself, not for economic or social ends, or even for professional advancement and recognition, although they are not indifferent to these.' Roe (1951, p. 65). Eiduson (1962, pp. 162 and 178 et seq.) makes a similar report based on her study of forty scientists. See also Darwin (1958, p. 141) for another scientist's disavowal of desire for recognition.

and, at times, to intense controversy. A historical summary and analysis of priority controversies has been given by Merton (1957), who pointed out that the failure to recognize previous work threatens the system of incentives in science. The pattern is not infrequent today. Of my seventy-nine informants, at least nine admitted to having been involved in questions of disputed priority either as the culprit or the victim. (The question was not asked in all seventy-nine interviews.) For example, an eminent theoretical physicist testified as follows:

> [This priority dispute] happened through an unfortunate habit of mine of not publishing things, delaying a year or two, talking about them, but not publishing. Therefore it has not always been clear to me whether something was known as my work or not. Under those conditions it is rather easy to get in priority disputes. It is much more satisfactory if one simply publishes what one does as he goes along. But a feeling of perfectionism frequently interferes with that. You don't want to publish something that's wrong. So you do the work, think about it for a couple of years, talk about it to everyone, but don't publish it. By that time somebody else has thought about it, and of course it's impossible to tell whether there was any influence from your work or not. It's a situation that's not very good. ... *Did this result in some hostility against you?*[3] Yes. One or two people – there haven't been many – were quite acrimonious. *They saw you about it?* Yes. ... It was a completely personal matter, and unfortunate.[4]

Another man, a little-known experimental physicist, was the victim in such a situation. A departmental colleague told about it first, when he was asked about the consequences of failure to recognize work:

> It causes a fatalistic attitude. ... A professor on our faculty concerned himself with the X effect long before it became popular in the early nineteen fifties. [He did his work in the late nineteen thirties.] His work

3. Italics in quotations are the interviewer's questions or comments. Statements in brackets are the writer's paraphrases or additional comments.

4. This may be a common failing of eminent men. Infeld wrote (1941, p. 277), regarding Einstein; 'Before we published our paper I suggested to Einstein that I should look up the literature to quote scientists who had worked on this subject before. Laughing loudly, he said: 'Oh yes. Do it by all means. Already I have sinned too often in this respect.'

was referred to *once* by a review from a large laboratory, and in the review his name was omitted – 'a professor at Y university' was the way they expressed it. This was deplorable. No action was ever taken on that. The man was just disappointed.

The experimental physicist himself described the same sequence of events, omitting the specific failure to recognize his accomplishments. Something similar to this had happened earlier in his career, when a grant he had requested was rejected, and shortly afterward someone else had become famous for doing essentially what he had proposed to do. This scientist was the most secretive man interviewed and was relatively isolated from his colleagues. In this case the individual affronted took no action; he may have reacted with hostility, but this was not communicated to the offenders. In other cases, as noted, affronted parties will communicate directly with the offenders. This may lead to a public recognition of priority – a 'Notice of Priority,' as it is called in the mathematical periodicals – or to an expression of recognition in succeeding papers by the offender or his collaborators. Thus, a mathematical statistician said that he was once anticipated in 'a serious matter':

...I hastened to recognize the other. In 1934... I overlooked a paper by a Russian, published twelve years before in an Italian journal. It was a very messy paper. In 1950 somebody from Iowa wrote me and informed me of these results, so I wrote a note of 'Recognition of Priority'.

The desire to obtain recognition induces scientists to publish their results. 'Writing up results' is considered to be one of the less pleasant aspects of research – it is not intrinsically gratifying in the way that other stages of a research project are. Some respondents were asked about the source of the greatest gratifications in research work. Generally the response was that most gratification came when the problem was essentially solved – when one became confident that an experiment would be successful, or when the outlines of a mathematical proof became clear – although details might remain to be cleaned up. For example, a theoretical physicist said: '[I get most pleasure] when the problem has been solved in principle but when some hard work remains to be done – when you have enough security to know you're not wasting your time but while there is some challenge left in the problem.'

An experimental physicist said one receives most gratification: '... when you find the effect you're looking for – everything else is anticlimax. Also in seeing some new and unexpected effect – in seeing new phenomena before others do.'

Research is in many ways a kind of game, a puzzle-solving operation in which the solution of the puzzle is its own reward.[5] 'Everything else' – including the communication of results – 'is anti-climax.' Writing up results, 'cleaning up loose ends', may be an irksome chore. A mathematical statistician said he was often pleased to discover that the result he had obtained was already in the literature:

Being anticipated doesn't bother me. Maybe it does once in a while, when I have something really nice. But if someone else has published the result it means I don't have to write it out, and that is gratifying. The real pleasure in the work comes in working the problem out. Publishing is necessary for money and is nice in a way. I try to publish what I find amusing and what I hope others will find amusing as well. ... Actually, I shouldn't give the impression that one publishes only in order to survive. People who are in really secure positions publish anyway. For example, [a very eminent man in the department] publishes about five papers annually, and he wouldn't have to publish at all any more.

The desire to obtain social recognition induces the scientist to conform to scientific norms by contributing his discoveries to the larger community. Thomas Sprat, writing near the dawn of modern science, perceived the importance of this: 'If neither *Chance*, nor *Friendship*, nor *Treachery* of Servants, have brought such Things out; yet we see *Ostentation* alone to be every day powerful enough to do it. This Desire of Glory, and to be counted Authors, prevails on all. ...'[6]

Not only does the desire for recognition induce the scientist to communicate his results; it also influences his selection of problems and methods. He will tend to select problems the solution of which

5. See Kuhn (1962, pp. 35–42). The layman can get some ideas of the gratification involved by reading such novels as C. P. Snow's *The Search* or Sinclair Lewis' *Arrowsmith*.

6. Sprat (1673, pp. 74 et. seq.). See also Karl Mannheim (1952, ch. 6 especially pp. 239, 242–3, 272) on the importance of the desire for recognition in science and other cultural pursuits.

will result in greater recognition, and he will tend to select methods that will make his work acceptable to his colleagues.

The range of acceptable methods varies. In mathematics, for example, the standards of rigor have changed steadily. Bell (1945, p. 153) has pointed this out in noting that eighteenth-century mathematicians were lucky, by later standards, not to have made more mistakes than they did:

How did the master analysts of the 18th century – the Bernouilles, Euler, Lagrange, Laplace – contrive to get consistently right results in by far the greater part of their work in both pure and applied mathematics? What these great mathematicians mistook for valid reasoning at the very beginning of the calculus is now universally regarded as unsound.

In this field, and most others, the change of standards is one of progress; the later standards can be shown to be technically superior. However, the definition of appropriate standards is not a technical matter only. For example, one informant was relatively famous for a method he had devised for making a certain kind of biochemical analysis. This method depended on distinctive nutritional requirements of certain bacteria. He noted that his method, while clearly superior to its alternatives for some purposes, and while very widely used, tended to be neglected by chemists:

[The method] fell into some disrepute because we're using organisms, and the chemists said 'Huh. Nobody can do the quantitative work with an organism.' ... There are other methods which chemists adopt. Chemists are a peculiar breed. They feel it is slightly debasing to use organisms. They just couldn't do that. ... Most chemists use [other methods] for psychological reasons if nothing else. But I would never agree that other methods are generally better, except for specific purposes.

That is, social recognition in biochemical work done by chemists induces them to use techniques defined as distinctively chemical.[7]

7. Compare the reception by chemists of chromatographic techniques, discovered by Tswett, a Russian botanist, in 1906: '... the chromatographic method got off to a bad start ... for a lowly botanist to assault thus the whole chemical profession was unthinkable! ... the chromatographic method fell largely into disrepute, and ... Tswett spent the later part of his life in misery and poverty.' The importance of the technique was only recognized in the late nineteen twenties. (Meinhard, 1949.)

Similarly, in mathematics the style of a proof, its 'elegance', is often considered as important to its merit as the truth of the theorem proved. While there are technical reasons for this, there are distinctively social ones as well. A mathematician described a mixed case:

Let me tell you a story about a famous problem in topology called the Poincaré conjecture. Various proofs were talked about, maybe they were published, but errors were found in all of them. Then, three years ago a Japanese mathematician published a hundred-page proof of it. A hundred-page proof is very unusual. Nobody has gone through it yet and found an error – this would be very difficult – but nobody believes in it. The man didn't have any new methods, and people didn't think he could do it with what he had.

Conformity with methodological standards is necessary if social recognition is to be given for contributions.

Similarly, the goals of science as they are specified in particular disciplines at particular times cover a restricted range, and the process of the reward of social recognition tends to produce individual conformity to the differentiated goals. As Baker (1945, p. 33) has put it:

The scientist is able to construct a sort of scale of scientific values and to decide that one thing or theory is relatively trivial and another relatively important, quite apart from any question of practical applications. There is, as Poincaré has well said, 'une hiérarchie des faits'. Most scientists will agree that certain discoveries or propositions are more important because more widely significant than others, though around any particular level on the scale of values there may be disagreement.

Erwin Schrödinger, the eminent theoretical physicist, has indicated (1935, pp. 76–80) that this has been true in his experience as well:

... it might be said that scientists all the world over are fairly-well agreed as to what further investigations in their respective branches of study would be appreciated or not. ... The argument applies to the research workers all the world over, but only of *one* branch of science and of *one* epoch. These men practically form a unit. It is a relatively small community, though widely scattered, and modern methods of communication have knit it into one. The members read the same periodicals. They exchange ideas with one another. And the result is that there is a fairly definite agreement as to what opinions are sound on this point or that. . . . In this respect international science is like international sport. . . . Just as it would be useless for some athlete in the world of sport to puzzle his brain in order to initiate something new – for he would have little or no hope of being able to 'put it over', as the saying is – so too it would, generally speaking, be a vain endeavor on the part of some scientist to strain his imaginative vision toward initiating a line of research hitherto not thought of.

Sanctions to enforce conformity in this respect, as well as with regard to appropriate techniques, are of two general kinds. First, works that deviate too far from the norm will be refused publication in scientific journals. A mathematician interviewed had a paper rejected for this reason:

A couple of us did some work that another person thought was merely trivial. And there was a little cat fighting going on here. . . . This was rejected by one journal, we felt for poor reasons; we rewrote it and sent it in again, and it was again rejected, we felt because of a sloppy refereeing job. I have been tempted and may yet write a complaining letter to the editor, whom I know, not asking that the paper be reconsidered but that the referee's work be evaluated, because the paper was not well refereed.[8]

Such an exercise of sanctions makes it impossible for the great mass of scientists to evaluate for themselves the importance and validity of the information presented. Delegating considerable power to a few authorities obviously infringes on the norms of independence in science. For this reason, editors and referees tend to be tolerant, basing their decisions on estimates that others will find manuscripts interesting even if they themselves do not. This and the fact that there are many journals, some of them unrefereed,[9] means that the sanction is of little importance to most scientists most of the time. A more important sanction is the

8. The comment of a referee of a biochemical journal: 'If the editor reversed what I said he would have to look for a new referee awfully fast, because I don't take that job lightly. And I've turned down many papers, too.'

9. In a survey by the International Council of Scientific Unions of one hundred fifty-six editors of well-known primary journals, 16 per cent reported that manuscripts were not sent to referees and 8 per cent gave equivocal answers to a question on the topic. (Porter, 1963.)

social recognition published work receives; this sanction is exercised by the community at large and applies to all published research.

Another type of sanction is not primarily important in science, although it is often alleged to be. This consists of extrinsic rewards, primarily position and money.[10] It is alleged that scientists publish, select problems, and select methods in order to maximize these rewards. University policies that base advancement and salary on quantity of publication sometimes seem to imply that this is true, that scientists' research contributions are not freely given gifts at all but are, instead, services in return for salary. While it is important for extrinsic rewards to be more or less consistent with recognition, the ideal seems to be that they should follow recognition, and this seems to be the general practice. In any case, an explanation of scientific behavior in terms of extrinsic rewards is weakened by the fact that many scientists in elite positions, whose extrinsic rewards will be unaffected by their behavior, continue to be highly productive and to conform to scientific goals and norms. Furthermore, scientists usually feel that it is degrading and improper to submit manuscripts for publication primarily to gain position without really caring if the work is read by others.

But why should gift-giving be important in science when it is essentially obsolete as a form of exchange in most other areas of modern life, especially the most distinctly 'civilized' areas? Gift-giving, because it tends to create particularistic obligations, usually reduces the rationality of economic action. Rationality is maximized when 'costs' of alternative courses of action can be assessed, and such costs are usually established in free-market exchanges or in the plans of central directing agencies. When participants are paid a money wage or salary for their efforts, and when this effectively controls their behavior, the system is more

10. Marcson (1960, p. 73) shows how organizations 'structure recognition by means of appropriate symbols, including titles, size of office, accessibility, financial rewards, and so on'. However, the difference between recognition and other rewards should be kept clear, for otherwise it is difficult to analyze social control and to specify the source of control. In this work 'recognition' means only the written and verbal behavior and the 'expressive gestures' of scientists that indicate their approval and esteem of a colleague because of his research accomplishments.

flexible than when controls derive from traditional or gift obligations.[11] Why, then, does this frequently inefficient and irrational form of control persist in science? To be sure, it also tends to persist in other professions. Professionals are expected to be motivated by a desire to serve others (Parsons, 1954, ch. 2). For example, physicians do receive a fee for service, yet they are expected to have a 'sliding scale' and serve the indigent at reduced fees or for no fees at all. The larger community recognizes two types of public dependence on professions: professional services are regarded as essential, concerned with values that should be realized regardless of a client's ability to pay; and nonprofessionals are unable to evaluate professional services, which makes them vulnerable to exploitation by unqualified persons. The rationale for the norm of service is usually the former type of dependence. In science, for example, the fact that a community has no one willing and able to pay for an important item of useless knowledge is not supposed to interfere with its ability to acquire the knowledge. But the idea of the gift and the norm of service is also related to the dependence of the public that follows from its inability to evaluate services.

The rationality of professional services is not the same as the rationality of the market. In contractual exchanges, when services are rewarded on a direct financial or barter basis, the client abdicates, to a considerable degree, his *moral control* over the producer. In return, the client is freed from personal ties with the producer, and is able to choose rationally between alternative sources of supply. In the professions, and especially in science, the abdication of moral control would disrupt the system. The producer of professional services must be strongly committed to higher values. He must be responsible for his products, and it is fitting that he be not alienated from them. The scientist, for example, must be concerned with maintaining and correcting existing theories in his field, and his work should be oriented to this end. The exchange of gifts for recognition tends to maintain such orientations. On the one hand, the recipient of the gifts finds it difficult to refuse them

11. Put another way, flexibility and rationality are maximized when workers are alienated from their products. See Parsons (1959, p. 89). See also Weber's stress on 'formally free' and actually alienated labor as a defining characteristic of capitalism (1927, p. 277).

(they are 'free'), and, on the other, the donor is held responsible for adhering to central norms and values. The maxim, *caveat emptor*, is inapplicable.[12] Furthermore, the donor is not alienated from his gift, but retains a lasting interest in it. It is, in a sense, his property.[13] One indication of this is the frequent practice of eponymy, the affixing of the name of the scientist to all or part of what he has found.[14]

Emphasis on gifts and services occurs frequently in social life, and we can get at the root of this generality by focusing on certain paradoxical elements implicit in the argument presented thus far. It has been argued that scientists are oriented to receiving recognition from colleagues and that this orientation influences their research decisions. Yet evidence that scientists themselves deny this has also been presented. There is a normative component to this denial, one that appears more clearly in analyzing scientific fashions. It is felt that, if a scientist's decisions are influenced by the probability of being recognized, he will tend to deviate from certain central scientific norms – he will fail to be original and critical. Thus, while it is true that scientists are motivated by a desire to obtain social recognition, and while it is true that only work on certain types of problems and with certain techniques will receive recognition at any particular time, it is also true that, if a

12. This does not mean that scientists are not supposed to be skeptical; that they should be skeptical about their own work as well as that of their colleagues is one of the more important institutionalized norms of science. (See Merton, 1949, pp. 315 et seq.) It does mean that unlike the consumer in the free market, the 'consumer' of scientific products can hold the producer morally responsible for 'defective products'.

13. See Merton (1949, pp. 312 et seq., and 1957 pp. 640 et seq.). Mauss (1954, pp. 41 et seq., et passim) pointed out that members of some 'archaic societies' felt gifts somehow remained part of the donor and that this belief was reinforced by further existential beliefs, e.g., the gift itself possessed the power to harm the recipient if it was not reciprocated.

14. See Merton (1957 pp. 642–4). See also Boring (1963, pp. 5–28), where it is suggested that eponymy has psychological benefits even for scientists who cannot hope to be remembered this way themselves. Compare Gall's questionnaire study of medical scientists (1960) in which he attempted to determine the extent to which they favored or opposed the use of eponymous terms. His results were inconclusive, which is not surprising since incentives of this type are seldom adopted on rational grounds or evaluated according to technical standards.

scientist were to admit being influenced in his choices of problems and techniques by the probability of being recognized, he would be considered deviant. That is, if scientists conform to norms about problems and techniques as a result of this specific form of social control, they are thereby deviants.

This apparent paradox, that people deviate in the very act of conforming, is common whenever people are expected to be strongly committed to values. In general, *whenever strong commitments to values are expected, the rational calculation of punishments and rewards is regarded as an improper basis for making decisions.* Citizens who refrain from treason merely because it is against the law are, by that fact, of questionable loyalty; parents who refrain from incest merely because of fear of community reaction are, by that fact, unfit for parenthood; and scientists who select problems merely because they feel that in dealing with them they will receive greater recognition from colleagues are, by that fact, not 'good' scientists. In all such cases the sanctions are of no obvious value: they evidently do not work for the deviants, and none of those who conform admit to being influenced by them. But this does not mean that the sanctions are of no importance; it does mean that more than overt conformity to norms is demanded, that inner conformity is regarded as equally, or more, important.

Thus, the gift exchange (or the norm of service), as opposed to barter or contractual exchange, is particularly well suited to social systems in which great reliance is placed on the ability of well-socialized persons to operate independently of formal controls. The prolonged and intensive socialization scientists experience is reinforced and complemented by their practice of the exchange of information for recognition. The socialization experience produces scientists who are strongly committed to the values of science and who need the esteem and approval of their peers. The reward of recognition for information reinforces this commitment but also makes it flexible. Recognition is given for kinds of contributions the scientific community finds valuable, and different kinds of contributions will be found valuable at different times.

The scientist's denial of recognition as an important incentive has other consequences related to those already mentioned. When peers exchange gifts, the denial of the expectation of reciprocity in kind implies an expectation of gratitude, a highly diffuse response.[15] It will be shown that this kind of gift exchange occurs among scientists, although the more important form of scientific contribution is directed to the larger scientific community.[16] In this case the denial of the pursuit of recognition serves to emphasize the universality of scientific standards: it is not a particular group of colleagues at a particular time that should be addressed, but all possible colleagues at all possible periods. These sentiments were expressed with his typical fervor by Johannes Kepler:

I have robbed the golden vessels of the Egyptians to make out of them a tabernacle for my God, far from the frontiers of Egypt. If you forgive me, I shall rejoice. If you are angry, I shall bear it. Behold, I have cast the dice, and I am writing a book either for my contemporaries, or for posterity. It is all the same to me. It may wait a hundred years for a reader, since God has also waited six thousand years for a witness.[17]

While this orientation is consistent with the scientist's need for autonomy – being dependent on the favors of particular others is terrifying – it also contains a strong element of the tragic. Scientists learn to *expect* injustice, the inequitable allocation of rewards. Occasionally one of them makes this explicit. Max Weber addressed students on 'Science as a Vocation' in the following way:

I know of hardly any career on earth where chance plays such a role. . . . If the young scholar asks for my advice with regard to habilitation, the responsibility of encouraging him can hardly be borne . . . one must ask every . . . man: Do you in all conscience believe that you can stand seeing mediocrity after mediocrity, year after year, climb beyond you, without becoming embittered and without coming to grief? Naturally, one always receives the answer: 'Of course, I live only for my calling.' Yet, I have found that only a few men could endure this situation without coming to grief.[18]

15. As Simmel says (1950, p. 387), gratitude 'establishes the bond of interaction, of the reciprocity of service and return service, even where they are not guaranteed by external coercion.' He goes on to note that persons make great efforts to avoid receiving gifts in order not to make such commitments to others. Something like this may occur in science.

16. In other words, the scientific contribution is closely approximated by the sacrificial model of the gift. See Durkheim (1915, pp. 342 et seq.): 'The sacrifice is partially a communion; but it is also, and no less essentially, a gift and an act of renouncement.'

17. Quoted in Koestler (1959, pp. 393 et seq.)

18. Gerth and Wright Mills (1946, pp. 132, 134). Weber was partly

More common than such an explicit statement is the myth of the hero who is recognized only after his death. This myth is important in science, as in art, because it strengthens universal standards against tendencies to become dependent on particular communities. Thus, mathematics has such heroes as Galois, who wrote the major part of his great mathematical opus the night before he was killed in a duel at the age of twenty-one; Abel, who died of tuberculosis as his greatness was coming to be recognized; and Cantor, who died mad, believing his ideas were spurned by others.[19] The stories of Copernicus, receiving his revolutionary book the day he died, and Mendel, rediscovered years after his death, are well known in the larger society, where they perhaps serve a function similar to that performed in science, albeit more general.[20]

References

BAKER, J. R. (1945), *Science and the Planned State*, Allen & Unwin.

BARTON, H. A. (1963), 'The publication charge plan in physics journals', *Physics Today*, vol. 16, no. 6, pp. 45–57.

BELL, E. T. (1937), *Men of Mathematics*, Simon & Schuster.

BELL, E. T. (1945), *The Development of Mathematics*, McGraw–Hill, 2nd edn.

BORING, E. G. (1963), 'Eponym as placebo', in E. G. Boring (ed.), *History, Psychology and Science: Selected Papers*, Wiley.

DARWIN, C. (1958), *The Autobiography of Charles Darwin, 1809–1882*, Nora Barlow (ed.), Collins.

DURKHEIM, E. (1915), *The Elementary Forms of Religious Life*, trans. J. W. Swain, Allen & Unwin.

EIDUSON, B. T. (1962), *Scientists: Their Psychological World*, Basic Books.

GALL, E. A. (1960), 'The medical eponym', *Amer. Scientist*, vol. 48, pp. 51–70.

GERTH, H. H., and WRIGHT MILLS, C. (1946) (eds. and trans.), *From Max Weber: Essays in Sociology*, Oxford University Press, New York.

GOULDNER, A. W. (1960), 'The norm of reciprocity', *Amer. sociol. Rev.* vol. 25, pp. 161–78.

---

concerned with the particular aspects of science in German universities, but the entire essay shows that he was also concerned with the more universal aspects of science as a profession.

19. See Bell (1937), for these and others.

20. The story of Job is the classic version of the hero who serves without reward.

Infeld, L. (1941), *Quest: The Evolution of a Scientist*, Doubleday.
Koestler, A. (1959), *The Sleepwalkers*, Hutchinson.
Kuhn, T. S. (1962), *The Structure of Scientific Revolutions*, University of Chicago Press.
Mannheim, K. (1952), *Essays on the Sociology of Knowledge*, Routledge & Kegan Paul.
Marcson, S. (1960), *The Scientist in Industry*, Harper & Row.
Mauss, M. (1954), *The Gift: Forms and Functions of Exchange in Primitive Societies*, Free Press.
Meinhard, J. E. (1949), 'Chromatography: a perspective', *Science*, vol. 110, pp. 387–920.
Merton, R. K. (1949), *Social Theory and Social Structure*, Free Press.
Merton, R. K. (1957), 'Priorities in scientific discovery', *Amer. sociol. Rev.*, vol. 22, pp. 635–59.
Parsons, T. (1959), 'Voting and the equilibrium of the American political system', in E. Burdick and A. J. Brodbeck (eds.), *American Voting Behaviour*, Free Press.
Porter, J. R. (1963), 'Challenge to editors of scientific journals', *Science*, vol. 141, pp. 1014–1017.
Roe, A. (1951), 'A psychological study of eminent biologists', *Psychol. Monogr.*, no. 65, p. 65.
Schrodinger, E. (1935), *Science and the Human Temperament*, trans. James Murphy, Allen and Unwin.
Simmel, G. (1950), *The Sociology of Georg Simmel*, K. Wolff, (ed. and trans.), Free Press.
Smelser, N. J. (1959), 'A comparative view of exchange systems', *Economic Development and Cultural Change*, vol. 7, pp. 173–82.
Sprat, T. (1673), *The History of the Royal Society of London*, London.
Weber, M. (1927), *General Economic Theory*, Free Press.

6 W. O. Hagstrom

The Differentiation of Disciplines

Excerpt from W. O Hagstrom, *The Scientific Community*, Basic Books, 1965, pp. 222–6.

Segmentation begins with cultural change, the appearance of new goals in the scientific community. Of course, new goals do not spontaneously 'appear': scientists actively seek them. Those who discover important problems upon which few others are engaged are less likely to be anticipated and more likely to be rewarded with recognition. Thus scientists tend to disperse themselves over the range of possible problems.[1] The behavior is analogous to competitive behavior in the animal world:

> But the struggle will almost invariably be most severe between the individuals of the same species, for they frequent the same districts, require the same food, and are exposed to the same dangers. In the case of varieties of the same species, the struggle will generally be almost equally severe, and we sometimes see the contest soon decided. ... (Darwin, 1958, p. 82.)

In many disciplines, dispersion to avoid competition takes place not only over the range of problems available but over the range of institutions, with the result that few identical specialists will be found in the same organizations. 'Complete competitors cannot co-exist.' (Hardin, 1960.)

Dispersion may lead to isolation, both geographical and social. Scientists working on the most unusual research problems, not being encouraged elsewhere, may be concentrated in a few research establishments. Social isolation results when pursuit of different goals leads to the development of different terminologies, techniques, and modes of organization. Eventually communication between specialities may be difficult and uncommon. Research in one area will have only remote effects on research in another, and programs of instruction may become differentiated;

1. Compare Durkheim's concept of moral density (1947, pp. 256–64).

the social equivalent of crossbreeding will occur less frequently:

Isolation, also, is an important element in the modification of species through natural selection. In a confined or isolated area, if not very large, the organic and inorganic conditions of life will generally be al-almost uniform; so that natural selection will tend to modify all the varying individuals of the same species in the same manner. Inter-crossing with the inhabitants of the surrounding districts will, also, be thus prevented. (Darwin, 1958, p. 106.)

Dispersion and isolation may encourage cultural differentiation, but here the biological metaphor breaks down. Differences between specialities may be viewed as deviance by members of specialties that are traditional or central to the discipline, and attempts may be made to sanction such deviance.

Initially, attempts may be made to establish conformity by the use of formal sanctions – with regard to appointments, instruction of students, and access to communication channels. The exercise of these sanctions tends to be implicit. Unsuccessful candidates for jobs may not be told specifically why they were not appointed, papers may be rejected by journals for vague reasons, and a professor may find his students failing because of their 'incompetence', or their 'attitudes'. Whether or not the recipient of the sanction is informed of the standards used in judging him, the matter does not become public.

Further development of the deviant specialty leads to overt social conflict. Those likely to be sanctioned publicly question the legitimacy of the standards used. Thus goals and standards are made explicit, and scientists and others will be made aware of the conflict. At first, organizations may attempt to reduce the resulting strains by primary adjustments. These usually amount to a limited range of permissiveness for the allegedly deviant group; they will be permitted a maximum number of appointments, limited access to channels of communication, and a higher degree of autonomy in instructing their own advanced students. Primary adjustments may make it possible to cope with conflict, but sometimes specialties will continue to diverge from one another, and the disadvantages of primary adjustments may lead to continued dissatisfaction. Such organizations as university departments and scientific societies will be made more rigid, scientific standards usually expressed by the award of recognition may become relaxed, and programs of instruction may suffer. Dissatisfaction with these failings may stimulate formal differentiation of disciplines.

Such differentiation requires special communication channels, the development of a disciplinary utopia, and successful appeals outside of existing disciplines. Leadership of an unusual sort in science – leadership unwilling to shrink from organizational controversy – will be necessary if these steps are to be successful. The establishment of communication channels and the development of a utopia make it possible for scientists to identify with the emerging discipline and to claim legitimacy for their point of view when appealing to university bodies or groups in the larger society.

At first, organizations may respond to these claims by structural innovations to which they need not be committed, innovations which they may abandon. In the university setting, this means such things as research institutes and interdisciplinary teaching programs, but not departments of instruction. Later on, separate departments for the new discipline may be established. This represents an almost irreversible differentiation, for universities are strongly committed to departments of instruction, and, through their graduates, the departments have reproduced themselves and established ties with groups in the larger society that employ them.

Structural change, especially departmental differentiation, is unlikely to occur unless the emerging discipline is marginal to at least two existing disciplines. When the emerging discipline is confined to a single existing discipline, it is apparently difficult for it to legitimate appeals outside of the disciplinary structure, especially appeals for structural change.

This analysis of social change can be conveniently presented in the schematic form in Figure 1.[2]

Differentiation results in the re-establishment of disciplines and specialties as the basic communities in science. New disciplines have internal structures similar to those from which they have differentiated, and this internal organization is characterized by the relatively great importance of informal relations.

2. This scheme would seem to be applicable to other types of segmentation in communities, such as the segmentation of religious groups, families, and nations.

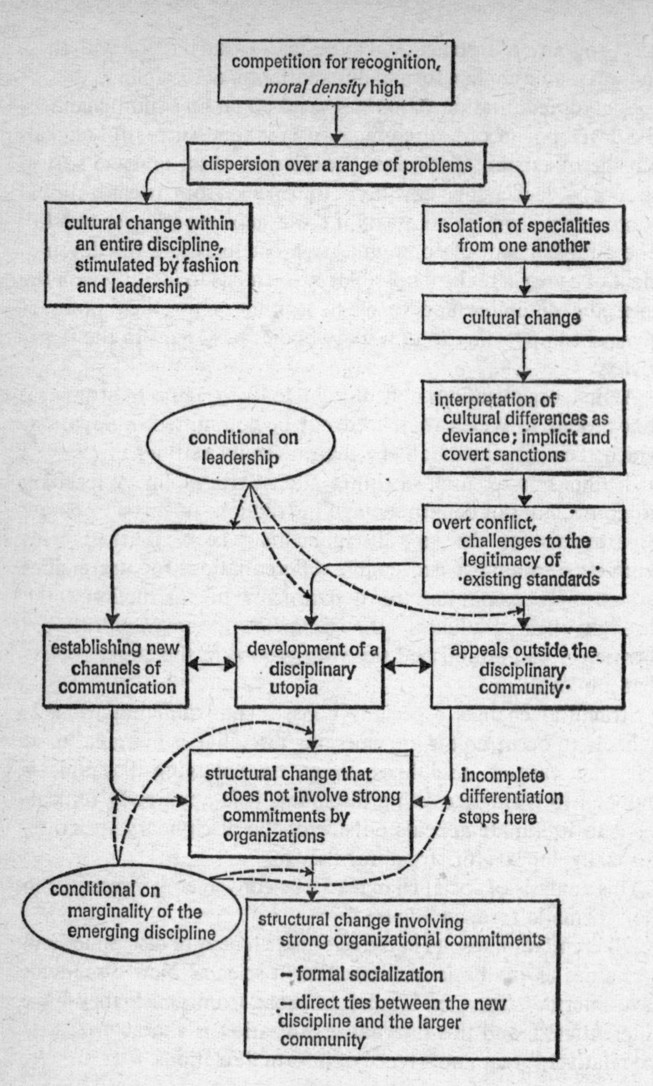

Figure 1 Differentiation in science: segmentation

From the point of view of the larger scientific community, however, continuing structural change implies a qualitative change in the organization of science. Modern science is no longer like the science of the seventeenth and eighteenth centuries. There are no universal scholars, and interdependence between disciplines and even among specialties within disciplines is small. This degree of specialization has implications for the organization of science that are reflected in the current problems of organizing university instruction and in allocating research facilities among disciplines. It also has implications for the place of science in the larger culture. Science no longer presents a unitary picture of the world to the nonspecialist, and the place of the scientist as a cultural leader is thereby dubious. This is expressed in a somewhat disillusioned passage written by Oppenheimer (1958):

> Today [as opposed to Plato's Greece], it is not only that our kings do not know mathematics, but our philosophers do not know mathematics and – to go a step further – our mathematicians do not know mathematics. Each of them knows a branch of the subject and they listen to each other with a fraternal and honest respect; and here and there you find a knitting together of the different fields of mathematical specialization. ... We so refine what we think, we so change the meaning of words, we build up so distinctive a tradition, that scientific knowledge today is not an enrichment of the general culture. It is, on the contrary, the possession of countless highly specialized communities who love it, would like to share it, and who make some efforts to communicate it; but it is not part of the common human understanding. ... We have in common the simple ways in which we have learned to live and talk and work together. Out of this have grown the specialized disciplines like the fingers of the hand, united in origin but no longer in contact.[3]

References

DARWIN, C. (1958), The Origin of Species, *Mentor*, 6th edn.

DURKHEIM, E. (1947), *The Division of Labour in Society*, trans. G. Simpson Free Press.

HARDIN, G. (1960), 'The competitive exclusion principle', *Science*, vol. 131, pp. 1292–7.

OPPENHEIMER, R. (1954), *Science and the Common Understanding*, Simon & Schuster.

OPPENHEIMER, R. (1958), 'The tree of knowledge', *Harper's*, vol. 217, pp. 55–57.

3. See also Oppenheimer (1954).

7 Michael Mulkay

Cultural Growth in Science[1]

Abridgement of M. Mulkay, 'Some aspects of cultural growth in the natural sciences', *Social Research*, vol. 36, no. 1, 1969.

Current sociological analysis of the processes involved in scientific development is couched predominantly in functionalist terms and derives primarily from the work of Robert Merton.[2] Although Merton's approach has come to be adopted by sociologists almost without question it has a very tenuous empirical foundation and involves considerable theoretical difficulties. [. . .]

I would like to demonstrate the inadequacy of this functionalist view by showing how scientific theory and methodological rules operate as *the dominant source of normative controls* in science and, in fact, as a basic hindrance to the development and acceptance of new conceptions.[3] It is, of course, at present impossible to provide detailed evidence taken from a large sample of instances showing that commitment within science to a specific body of knowledge and its associated techniques is greater than commitment to any purely social norms. What I shall do, therefore, is to examine an extreme case where marked deviation from the Mertonian norms

1. I must at the outset clarify two points. First, the term 'culture' is used here to refer to the sphere of cognitive symbols in science, i.e., scientific theories, conceptual schemes, methodology and techniques of inference. In this sense the culture of science is quite distinct from the social structure of science. I am concerned in this paper primarily with the cultural development which, within modern science, tends to precede structural change. Secondly, the analysis does not apply to pre-seventeenth-century science.

[The greater part of the abridged passages here summarises the work of Merton and of Kuhn. The earlier passages by these authors provide an adequate substitute for this material. Ed.]

2. For example, Barber (1962); Storer (1966); Stein (1962).

3. It will be suggested later that established theories, etc., operate as a barrier to innovation only under specific conditions and that built-in resistance to change does in fact facilitate certain forms of scientific development.

and radical theoretical innovation occur together. In this way we shall be able to compare the consequences of deviation from social norms with those arising from the presentation of radical theoretical innovations. It is generally accepted that deviation from established norms generates corrective responses within the group concerned. (See for example Homans, 1950, ch. 11.) One of these responses tends to be the continued re-statement of the norm itself as a means of bringing the recalcitrant back into line.

[For] as we know from the sociological theory of institutions, the expression of disinterested moral indignation is a signpost announcing the violation of a social norm. Although the indignant bystanders are themselves not injured by what they take to be the misbehaviour of the culprit, they respond with hostility and want to see 'fair play', to see that behaviour conforms to the rules of the game. (See Merton, 1962, p. 453.)

Situations involving social deviance, then, are useful for the sociologist in that they bring into the open normative commitments which might otherwise remain implicit. They also provide a rough means of measuring the relative importance of specific norms within the group. I shall therefore examine briefly the most massive case of theoretical, methodological and 'social' non-conformity in the recent history of science.

The case of Dr Velikovsky

In 1950 Dr Immanuel Velikovsky published a book called *Worlds in Collision*, which challenged many of the central presumptions of astronomy, geology and historical biology. (Velikovsky, 1950.) I am not interested here in the precise nature of Velikovsky's claims nor in their scientific validity, though the latter is relevant in some degree to the discussion which follows, but in the reception of his work by the scientific community. De Grazia (1963) describes the book's reception in these words:

The publishing industry, numerous scientists, the popular press, and a considerable public in America and abroad engaged in acrimonious debate and bellicose maneuvers over the validity of the new historical and cosmological concepts of Dr Immanuel Velikosvky.[4]

4. The discussion here is based primarily upon the information provided in *The American Behavioral Scientist* (de Grazia *et al.*, 1963). [Compare the approach in Polanyi, 1967. Ed.]

How can we explain this reaction to Velikovsky's work, not so much on the part of the general public, but by practicing scientists? One reason may be that Velikovsky undoubtedly broke the rule of communality at the outset by allowing popular interpretations of his work to be published before the main opus itself had been presented. But this relatively minor infringement of scientific norms cannot account for the violence of scientists' responses to Velikovsky's theses. I suggest that the turbulence which his hypotheses created among scientists was due to Velikovsky's rejection of those theoretical and methodological paradigms accepted at that time within a number of scientific disciplines. These paradigms operate as norms. Not only do they supply cognitive and perceptual frameworks but they also provide standards for judging the acceptability of hypotheses. Velikovsky's massive departure from the theoretical and methodological limits set by these paradigms led many scientists into an emotional rebuttal of his propositions and into a denial of his scientific integrity.

Numerous examples can be cited of violation by members of the scientific community of the Mertonian norms. In February, 1950, severe criticisms of Velikovsky's work were published in *Science News Letter* by experts in the fields of astronomy, geology, archaeology, anthropology and oriental studies. None of these critics had at that time seen *Worlds in Collision*, which was only just going to press. These denunciations were founded upon popularized versions published, for example, in *Harper's*, *Reader's Digest* and *Collier's*. The author of one of these articles, the astronomer Harlow Shapley (see de Grazia *et al.*, 1963), had earlier refused to read the manuscript of Velikovsky's book because Velikovsky's 'sensational claims' violated the laws of mechanics. Clearly the 'laws of mechanics' here operate as norms, departure from which cannot be tolerated. As a consequence of Velikovsky's non-conformity to these norms Shapley and others felt justified in abrogating the rules of universalism and organized skepticism. They judged the man instead of his work and in this way failed to live up to the demands of organized skepticism, for they did not subject Velikovsky's claims to rigorous examination before assessing the validity of these claims. I have drawn attention to Shapley's behavior because it represents an extreme instance of the way in which certain scientists reacted to Velikovsky, Shapley refused to read the manuscript; he refused to test Velikovsky's specific prediction with respect to the presence of hydrocarbon dust and gases in the atmosphere around Venus;[5] yet he actually wrote a damning critique of *Worlds in Collision* and advised others to do likewise. Although an extreme example, Shapley's reaction is typical of that of many scientists in the United States, the United Kingdom and elsewhere. In addition to violating, in a variety of ways, norms of universalism and organized skepticism, the scientific community also failed to allow its members unrestricted access to the information presented by Velikovsky. Thus the rule of communality was infringed. For example, Velikovsky attempted to answer his critics at a meeting of the American Philosophical Society in April, 1952, devoted specifically to 'Some Unorthodoxies of Modern Science'. Although Velikovsky's arguments received respectful attention and discussion during the meeting itself, his request that his remarks be printed in the society's *Proceedings* along with those of his adversaries was refused. The same disposition to restrict dissemination of Velikovsky's ideas can be seen in the movement which arose in favor of applying sanctions against the publishers of *Worlds in Collision.* Scientists wrote letters of complaint to Macmillan Company, declined to see its business representatives, and threatened to boycott its scientific textbooks. Not only did these actions violate norms of communality but they also infringed the individual scientist's right to make up his own mind as to the validity of Velikovsky's claims. Similarly, informal pressures were brought to bear upon those who suggested that these revolutionary ideas should at least be open for discussion. Gordon Atwater was dismissed from his positions as Curator of Hayden Planetarium and Chairman of the Department of Astronomy at the American Museum of Natural History. Atwater seems to have committed two offenses. In the first place, he was one of the outside readers who recommended to the Macmillan Company that *Worlds in Collision* should be printed. And secondly, he was preparing to present a planetarium program to the public portraying the astronomical events described in the book. This is a specific example, then, of scientists

5. As in so many cases Velikovsky's prediction has since been verified, in this instance by the tests carried out in association with Mariner II.

applying sanctions against one of their fellows for supporting Velikovsky, and thereby clearly disregarding the norm of scientific independence. It is also evident that this whole affair represents a marked rejection by sectors of the scientific community of the value of original thought. Originality is undoubtedly valued by scientists, but originality which remains within limits imposed by the existing 'style of thought'. If originality takes the form of 'a stupendous panorama of terrestrial and human histories which will stand as a challenge to scientists to frame a realistic picture of the cosmos' (see O'Neill, 1946), or any other form which requires possible reformulation of paradigms which have attained the support of a quasi-moral commitment, then originality is likely to be rejected without regard for the putative ethical restraints imposed by the scientific ethos.

This brief outline of the reception of Velikovsky's work would seem to indicate that the norms outlined earlier simply do not govern the behavior of scientists to any noticeable degree. However, almost from the beginning there was a reaction against the treatment accorded Velikovsky. This reaction took the form of a reaffirmation of scientific norms rather than a defense of Velikovsky's specific assertions. The reaffirmation of these norms was justified by the norm of disinterestedness, that is, that the actions of scientists should be directed to the benefit of the total scientific community. Conformity to scientific norms, it was claimed, was necessary in order to insure the vitality of science. Whatever the merits of Velikovsky's specific hypotheses, their abusive and uncritical rejection was harmful to the community at large and to the continued extension of certified knowledge. This reaction does provide some evidence pointing to the existence of Mertonian norms in the scientific community. However, what is particularly noticeable about the Velikovsky affair is not this mild and uncertain reaction against the widespread failure within the scientific community to conform to the Mertonian norms but the persistent tendency of scientists both in the United States and elsewhere, through the medium of published reviews as well as personal contacts, to justify rejection of Velikovsky's claims simply by indicating the latter's departure from established beliefs.

This outline of the Velikovsky case demonstrates a basic flaw in the functionalist analysis of the scientific community. According to the functionalist approach we should have predicted that Velikovsky's work would have been subjected to detailed and critical examination by those scientists directly concerned. If any scientists had reacted publicly in an emotional and negative manner they would have been restrained by reaffirmation of the Mertonian norms by their professional colleagues. As we have seen, this did not happen. Instead we find extensive deviation from the Mertonian norms in terms of actual behavior.[6] Furthermore, it is not the Mertonian norms which are affirmed as a means of subduing the socially deviant critics of Velikovsky; rather the established theoretical and methodological models are publicly affirmed as a means of subduing the recalcitrant Velikovsky and his 'supporters'. All this would seem to indicate that theoretical and methodological norms are more central to the structure of the scientific community than are the Mertonian social norms, at least when radical and thereby threatening innovations are involved. The Velikovsky case points to rigidity rather than fluidity in the intellectual commitment of the natural scientist. This rigidity becomes more intelligible when we recall the intensive educational process in science and that the *intended* goal of this process is primarily the attainment of familiarity with a body of established knowledge rather than the inculcation of social norms. We know that, for several reasons, socialization in science is likely to be particularly effective. First, it lasts well into adult life. Secondly, its severe demands upon the student's time and its highly specialized skills tend to isolate the participants from alternative vocational and intellectual interests. In the third place, science graduates tend to be particularly dependent upon their teachers for technical guidance and eventual employment. (See Berelson, 1960.) Furthermore, Kuhn (1963) suggests that the science students within each specialty are educated by learning systematically just one consistent approach to the problems they will be faced with as fully-fledged professionals. He concludes that 'nothing could be better calculated to produce "mental sets"' than the present system of education in the sciences. Hagstrom (1965, ch. 1) further suggests that 'the effects of scientific socialization are reinforced by a highly selective system of recruitment'. It is at

6. It seems likely that some of the Mertonian norms exist in science more as 'institutional fictions' at the verbal rather than the behavioral level.

least plausible to maintain that these processes of selection and socialization engender the kind of cognitive rigidity which has been postulated above to account for the reception of Velikovsky's work. We can in fact use the Velikovsky case once more to provide further support for the position adopted here if we assume that one product of the intensive socialization process in science is the emergence of a strong need for 'cognitive consensus'. Following Festinger (1957), 'cognitive consensus' refers to a logical consistency among cognitions and 'cognitive dissonance' to a logical inconsistency among cognitions. Festinger claims that cognitive dissonance serves as a source of motivation, 'as an antecedent condition which leads to activity oriented toward dissonance reduction'.[7] It is a reasonable inference that the need for cognitive consonance will be stronger in science as a consequence of the pattern of education than in most other communities. We should therefore be able to observe the operation of this need for cognitive consonance, especially in the reception of Velikovsky's revolutionary claims.

The magnitude of cognitive dissonance in a particular instance is a function of the importance of the elements involved. It is evident that the scientist's theoretical and methodological commitments are of central relevance for the continuance of his professional activity. Thus the intensity of scientific reaction against Velikovsky is in accord with the magnitude of the dissonance. Velikovsky not merely challenged the accepted view within one specialty; rather, he challenged basic theoretical assumptions in a number of disciplines and, by using historical records of natural events, he put in question the essential methodology of modern science. Once cognitive dissonance exists it can be reduced by changing one's established cognitions. But this is the more difficult the greater the degree of dissonance. Consequently, it was impossible for most scientists in this instance. (Of course, if norms of organized skepticism had been in operation scientists would have experienced little difficulty in putting Velikovsky's propositions

7. Festinger (1957, p. 3). I am in no way committed in this paper to Festinger's particular theoretical interpretation of cognitive dissonance. I need assume only that the educational experience of the scientist will generate a specially strong tendency towards reducing dissonance and that dissonance will be reduced along the lines suggested by Festinger.

to test without altering their essentially *conditional* acceptance of the established model.) If it is impossible to change one's own cognitive commitments there are, according to Festinger, only two alternatives open – to add new cognitive elements and to change 'environmental cognitive elements'. These alternatives are not mutually exclusive and scientists adopted them both. First, a new cognitive element was added by stressing that Velikovsky was not professionally qualified as a natural scientist and that, in addition, his integrity was suspect. As a result of this new cognitive element it became unnecessary to take Velikovsky's claims seriously and in this way cognitive dissonance was reduced. But primarily scientific activity was directed toward changing environmental elements and particularly toward preventing the dissemination of Velikovsky's propositions throughout the scientific world and the wider community. It is clear that widespread consideration of Velikovsky's claims, especially within the scientific community itself, would have generated extreme dissonance for many scientists. This possible development was avoided in a number of ways. Elaborate rebuttals were offered even before Velikovsky's work was published. Velikovsky was refused access to professional journals and thereby prevented from answering his critics. Some of his 'supporters' were dismissed from office. Organized sanctions were brought against his publishers. Scientists refused to test Velikovsky's predictions, even when this could have been done with little inconvenience. Finally, priority was not awarded to Velikovsky in the journals; his successful predictions were explained away as the luck which any charlatan is likely to enjoy. Most of the negative responses, then, towards Velikovsky served to reduce cognitive dissonance within a scientific community strongly committed to specific theoretical and methodological models which has been inculcated by an intensive socialization and recruitment process. It is clear that this analysis accords closely with all that has been stated above as to the normative role of the body of established knowledge within the scientific community.[8] However, if we accept that scientific 'models' operate as norms, that there exists a powerful strain towards

8. The strain towards cognitive consonance may, in varying degrees, operate to maintain conformity to established norms in all social communities.

cognitive consonance, and that scientific education fosters detailed knowledge of and commitment to one specific approach to the study of reality, then cultural change in science once more appears problematic. We must, therefore, look for an approach to this problem which provides an alternative to the functionalist view and which makes sense of the points established above.

An alternative to the functionalist approach

Adherents to the functionalist view could adapt to the criticisms offered so far by admitting that established theories have a conserving effect and that Mertonian norms are not overwhelmingly endorsed but by asserting, in addition, that growth only occurs to the extent that Mertonian norms do in fact operate. The underlying assumption here would be that growth can only take place within 'open' communities and that, as science has undoubtedly developed more rapidly than other intellectual movements, so the scientific community must be more 'open' than other social groups.[9] However, it is quite possible to view scientific growth as a product of intellectual and social closure. This is the basic assumption of Kuhn's assertion that scientific activity most of the time consists in the 'attempt to force nature into the conceptual boxes supplied by professional education'. [. . .]

The functionalist approach to the study of the social organization of research stresses that certified knowledge will accumulate automatically as a consequence of reasonable conformity to the Mertonian norms. Kuhn's view, however, emphasizes that scientists develop a strong commitment to a particular theoretical–methodological tradition and that there are consequently powerful forces within science working to limit the possibility and acceptance of innovations. Within Kuhn's view Mertonian norms need not be stressed in accounting for 'normal science'. Research will be guided by the accepted paradigm and professional recognition will be gained primarily by means of contributing to the further articulation of the paradigm within the limit provided by established technical and cognitive norms. In the case of 'revolutionary science' the Mertonian norms do not appear to operate

9. The assumptions of Merton and Barber as to the relations between 'openness' and scientific growth can be clearly discerned in their analyses of the link between science and democratic societies.

effectively, at least this is so if we can generalize from the case of Velikovsky. Furthermore, Kuhn has shown how paradigm-guided research is likely to produce rapid cumulation of knowledge by directing research within a strongly defined community to detailed consideration of those problems suggested by the paradigm. In addition, Kuhn claims that such detailed research, which would be unthinkable without the accepted paradigm, itself generates anomalies which lead to reconstruction of the existing model. In this sense modern science grows through the formation of closed specialties which generate new knowledge by restricting attention to a number of specific and solvable problems regarded as important within the group. However, Kuhn's scheme itself appears to have several deficiencies. In the first place, most of Kuhn's examples either deal with the emergence of specific disciplines from their pre-paradigmatic into their paradigmatic period, e.g., the work of Newton or Lavoisier, or are taken from theoretical physics. To quote Ben-David:

> This suggests that the feedback mechanisms involved in the exhaustion of paradigms and the generation of new ones might be quite efficient in overcoming organizational resistance in such deductive fields as mathematics and theoretical physics, where there might be little disagreement about the nature of puzzles and their number and duration at any time. But by the same token it suggests that basic ('revolutionary') changes in other fields require further organization mechanisms which are still to be identified. (Ben-David, 1964, p. 473.)

Secondly, as Holton (1962) has shown, a great deal of growth in science is due to the spread of paradigms into areas which either have not developed established paradigms of their own or which have not previously existed as distinct areas of inquiry. In most specific research areas there are a limited number of 'interesting ideas' available. The emergence of a new field generates a rapid influx of researchers and consequently a rapid development of these ideas. Within quite a short time, say five to ten years in the case of physics, no more basic contributions are being found or sought. Nevertheless, while the growth of each special area of research is limited in duration and scope the corpus of scientific knowledge itself continues to grow as new but related areas are opened up for investigation. 'When an important insight (including a "chance" discovery) causes a new branch line . . . to rise, fruitful research usually continues on the older line. . . . But many of the original people will transfer' to the new line 'and there put to work whatever is applicable from their experience' in the older area of research (Holton, 1962). Science tends to proceed therefore by means of discovery of new areas of ignorance. New areas of ignorance are not associated in the minds of scientists with established paradigms. As a consequence there is far less resistance to innovation. Let me give examples from Holton's own account (1962, p. 126).

[. . .] in 1895 Roentgen seemed to have exhausted all the major aspects of X-rays, but in 1921 the discovery of X-ray diffraction in crystals by von Laue, Friedrich, and Knipping transformed two separate fields, those of X-rays and of crystallography. Moseley in 1931 made another qualitative change by showing where to look for the explanation of X-ray spectra in terms of atomic structure, and so forth. Similarly the Joliot-Curie findings gave rise to work that had one branching point with Fermi, another with Hahn and Strassmann . . . Thus the growth of scientific research proceeds by the escalation of knowledge – or perhaps rather of new areas of ignorance – instead of by mere accumulation.

There is a multitude of instances where modern science has grown by the extension of paradigms established in one region into new areas. To some extent this is what happened at the level of the major scientific disciplines as well as at the level of the specialty. The developments in chemistry and biology during the seventeenth and eighteenth centuries were in part produced by the application of the methods and concepts found to have been successful in astronomy and physics during the sixteenth and seventeenth centuries. (See Butterfield, 1962, p. 146.) The same process of cross-fertilization characterizes the development of the social sciences. Comte's social physics and Durkheim's recourse to the methods of biology are obvious examples. Clearly this pattern of growth does not fit neatly into Kuhn's model where development takes the form of intensive research and eventual reformulation within established areas of study. However, if we combine these two views we have a scheme which makes sense of the intellectual inertia within science and the undoubtedly rapid growth of scientific knowledge. In certain fields which have clearly defined paradigms growth will be facilitated by systematic training, by group closure and by intensive research into a narrow range of problems. Within such fields major innovations will meet with strong resistance from the majority of scientists who have become committed to the existing paradigm. Less clearly defined and less deductive research areas will produce weaker commitment and change will therefore tend to be more gradual. At the same time, within all scientific disciplines there will be a strong tendency for ideas to escalate into new fields thereby facilitating the development of new knowledge without the accompaniment of vigorous resistance. This view of scientific development has several advantages. It centers the analysis around specific bodies of knowledge. It allows for the normative aspects of established paradigms. And it takes account of resistance to innovation while including the idea of rapid growth. In addition, it links up with Hagstrom's important work (1965, chs. 4 and 5) on the processes whereby scientific specialties become structurally differentiated. Where Hagstrom deals with the structural processes this paper concerns itself with the processes of 'cultural' growth[10] which, in science, precede structural differentiation.

Although the view of scientific development proposed above is preferable to the functionalist analysis in that it provides a more satisfactory account of the processes whereby ideas are *accepted* into science, little has been stated as yet with respect to the *generation* of ideas. Neither Merton nor Kuhn provides an analysis of the structural sources of innovation, yet this is surely a crucial aspect of scientific development. Holton, however, offers an important clue by stressing how ideas current in one area can escalate into totally new areas or can cross-fertilize previously unorganized fields. Cross-fertilization of ideas has been historically important in the growth of science and it is important today; although cross-fertilization today takes place within a relatively closed scientific community rather than between divergent scientific traditions. Some studies by Ben-David give us considerable insight into the mechanisms and structural determinants of cross-fertilization in modern science. These studies can also be linked systematically with the ideas presented so far in this paper.

10. Cultural in the sense defined above.

Cross-fertilization in modern science

In a comparative study of the growth of science in four societies Ben-David (1960a) shows that the rate of innovation is hindered by strongly centralized organization and assisted by a loose and competitive structure.[11] For example, nineteenth-century French science was centrally controlled while that of Germany was highly competitive owing to the absence of any governmental or other centralized controlling agency. As a consequence, German science became increasingly productive of new ideas during the nineteenth century while that of France fell steadily behind. Furthermore, as the loosely structured German system came to be adopted in the United States, that country's rate of innovation increased. This association between intellectual creativity and an open and competitive situation is not restricted to science. During the nineteenth century, as the central control over science in France became firmly established the centralized academy-system governing French painting was breaking down; and it is precisely as the academy-system became less and less effective that French painting produced its efflorescence of the second half of the nineteenth century. (See White and White, 1964.) It seems to me that the situation in painting is more complex than that in science; yet in both cases intellectual innovation appears to be positively related to the clash of ideas within an open social structure. It is reasonable to infer therefore that the underlying factor in both instances is what I have called above 'cross-fertilization', i.e., the interplay of divergent cognitive-normative frameworks.

Studies by Ben-David and by Barber serve to highlight another sociological facet of the process of generation of scientific ideas. Ben-David (1960b, p. 567) shows that in the emergence of bacteriology and psychoanalysis the innovators' interest in practical problems led them to new conceptions which were seen to be theoretically important.

> [...] analysis of the beginnings of bacteriology and psycho-analysis lends general support to the proposition that contact with practice may be important in reorienting research toward the investigation of new and fruitful problems. The practitioner-scientists appear as forerunners,

11. The scientific community can be 'open' in terms of its authority structure and yet 'closed' in terms of its commitment to cognitive norms.

supporters, and disciples in the history of the two innovations bacteriology and psychoanalysis, the central figures in both were 'role hybrids' who were led to the innovation by an abrupt change from theoretical research to applied science. The practitioner-scientist, as well as the role hybrid, each are in a position to shift frames of reference relatively easily.

Ben-David also supplies further evidence of the normative character of the established paradigms in these areas by documenting the resistance of scientists who on the whole were unwilling even to consider such radical departures from the accepted framework of their professional community. Ben-David has shown then that, in the field of medical research, practitioner-scientists and scientist-practitioners are more likely to produce and accept marked scientific innovations, as opposed to research findings 'expected' within the framework of the current paradigm, than are those researchers operating within a single frame of reference. It is clear that the relationship between occupancy of dual roles and innovation is closely related to the process of cross-fertilization which we have seen to be so important in science. We could perhaps therefore generalize Ben-David's finding and suggest that occupancy of dual roles, insofar as these roles entail distinct approaches to similar research problems, will tend to favor the generation of new cognitive frameworks. This proposition is, of course, no more than a tentative inference. It does however appear to me to deserve further investigation. Furthermore, there is an indication of how the process works out in practice in a study by Barber and Fox.

The study by Barber and Fox (1962) of The Case of the Floppy-Eared Rabbits is simply a comparison between the research activities of two scientists. Both scientists had accidentally collapsed rabbits' ears when injecting them intravenously with crude papain. One scientist, however, was unable to find an answer to the puzzle which this phenomenon presented. Barber and Fox suggest that this scientist did not experience 'positive serendipity' primarily because his research preconceptions led him to look for the wrong kind of answer. The floppiness of the rabbits' ears was due to 'quantitative change in the matrix of the cartilage' brought about by the papain. The first scientist missed this explanation because he had been taught to think of cartilage as relatively inert tissue and because his research aims centered around muscle tissue. In contrast, a second scientist did experience 'positive serendipity' in relation to the same phenomenon. However, this scientist was also initially directed away from the solution by his acceptance of the established paradigm in this area. The problem was only resolved when the researcher approached the problem from the angle of a teacher. While attempting to 'convey to students what experimental pathology is like' the biologist compared 'sections taken from the ears of rabbits which had been injected with papain with comparable sections from the ears of rabbits of the same age and size which had not received papain'. When taking the role of the teacher the scientist had no rigid conception of what kind of result should obtain. Indeed he states explicitly that when preparing this simple controlled comparison he did not expect to reach any conclusion. He was more concerned with demonstrating the method than with justifying his research preconceptions. It is clear that in this instance occupation of dual roles facilitated innovation by providing divergent frames of reference.[12]

At least a minimal diversity of roles, then, appears to promote scientific innovation. We have seen evidence for this in Ben-David's large-scale study of the relationship between scientific productivity and an open, competitive research structure; in his examination of how theoretical innovation in fields such as bacteriology benefits from role-duality; and in the study by Barber and Fox of the low level mechanics of innovation. We can, therefore, now summarize our tentative conclusions as to the nature of the 'generation process' along the following lines. All research is structured by existing paradigms or, in the pre-paradigm situation, by existing research findings. Within certain (deductive) disciplines in which there exist clearly defined problems, normal science generates in course of time sufficient anomalies to necessitate radical innovations and consequent redefinition of disciplinary problems. While normal science may be carried on quite effectively within centralized organizations, scientific innovation is facilitated by a fluid social structure which brings divergent frames of reference together. A great deal of the rapid expansion

12. There is a clear resemblance between the approach to innovation adopted here and the more psychological treatment offered by Koestler (1964).

of modern science has been due, not so much to the accumulation of systematically related knowledge within established areas, as to the more or less accidental discovery of new areas of ignorance. This kind of growth is particularly fast partly because there exist no trained and committed specialists to resist innovation in the new specialties and partly because paradigms from related areas can often be fruitfully applied. The actual genesis of new ideas tends to be associated with some form of cross-fertilization of ideas both at the level of detailed research and at the level of disciplinary paradigms. Within modern science cross-fertilization occurs primarily when individual researchers occupy dual or multiple roles related to their research. The relationship between innovation and occupation of dual roles exists because the established paradigm operates as a norm, i.e., it provides a frame of reference which guides intellectual performance and a standard for judging the legitimacy of research findings. Differing research roles provide divergent guides for intellectual activity and distinct judgements as to which findings are legitimate. Innovation occurs as the divergent cognitive frameworks are merged in some kind of synthesis.

References

BARBER, B. (1962), *Science and the Social Order*, Collier Macmillan.

BARBER, B. and FOX, R. (1962), 'The case of the floppy-eared rabbits', in B. Barber and W. Hirsch (eds.), *The Sociology of Science*, Free Press.

BEN-DAVID, J. (1960a), 'Roles and innovations in medicine', *Amer. sociol.*, vol. 65, pp. 557–68.

BEN-DAVID, J. (1960b), 'Scientific productivity and academic organization in nineteenth century medicine', *Amer. sociol. Rev.*, vol. 25, December, pp. 828–43.

BEN-DAVID, J. (1964), 'Scientific growth: a sociological view', *Minerva*, vol. 4, pp. 455–76.

BERELSON, B. (1960), *Graduate Education in the US*, McGraw–Hill.

BUTTERFIELD, H. (1962), *The Origins of Modern Science*, Collier Macmillan.

FESTINGER, L. (1957), *A Theory of Cognitive Dissonance*, Row, Peterson & Co.

GRAZIA, A. de (1963), JUERGENS, R. E., and STECCHINI, L. C. (1963), 'Dr. Velikovsky and the politics of science', *Amer. behav. Sci.*, vol. 7, September.

Hagstrom, W. O. (1965), *The Scientific Community*, Basic Books.
Holton, G. (1962), 'Models for understanding the growth and excellence of scientific research', in S. R. Graubard and G. Holton (eds.), *Excellence and Leadership in a Democracy*, Columbia University Press.
Homans, G. C. (1950), *The Human Group*, Harcourt, Brace & World.
Koestler, A. (1964), *The Act of Creation*, Hutchinson.
Kuhn, T. S. (1963), 'The essential tension, tradition and innovation in scientific research', in C. W. Taylor and F. Barron (eds.), *Scientific Creativity*, Wiley.
Merton, R. K. (1962), 'Priorities in scientific discovery', in B. Barber and W. Hirsch (eds.), *The Sociology of Science*, Free Press.
O'Neill, J. J. (1946), *New York Herald Tribune*, August 14.
Polanyi, M. (1967), 'The growth of science in society', *Minerva*, vol. 5, no. 4, pp. 533–45.
Stein, M. I. (1962), 'Creativity and the scientist', in B. Barber and W. Hirsch (eds.), *The Sociology of Science*, Free Press.
Storer, N. W. (1966), *The Social System of Science*, Holt, Rinehart & Winston.
Velikovsky, I. (1950), *Worlds in Collision*, Macmillan Co.
White, C., and White, H. (1964), 'Institutional change in the French painting world', in R. N. Wilson (ed.), *The Arts in Society*, Prentice-Hall.

Part Three **Science in Application: Links with Technology and Economy**

As science has grown, so the process of purification that produced disinterested academic research has been reversed; strong interrelationships have grown between scientific, technological and economic activities, so that today most scientists are employed in industrial and government laboratories. Scientific knowledge has become a necessity to modern developed economies, and whole industries have been based on innovations made possible by basic scientific advances. Consequently, science has enjoyed lavish economic support, and even the most esoteric and fundamental research areas have been free to expand as their own internal dynamics dictated. In many such areas, increasingly sophisticated basic theory has generated a need for more accurate and extensive empirical investigation, which has entailed more elaborate and costly technological aids, particle accelerators and cyclotrons, electron microscopes and radio telescopes. The result has been the growth of 'Big Science' in pure as well as applied areas.

'Big Science' involves teamwork, division of labour and complex allocations of rights and duties among large numbers of scientists; it generates new patterns of communication, erodes existing systems of individual recognition and creates a demand for scientists with administrative skill. When it grows within the institutional framework of academic research, a framework which evolved entirely in relation to the practice of 'Little Science', it constitutes a potent source of normative change. Thus, indirectly, the utilitarian successes of science have pervasively influenced its entire social organization, as well as having shifted its institutional locale.[1]

1. Weinberg (1970) has recently produced an excellent discussion of the growth of teamwork in science, its relationship to 'Big Science', and its

The paper by Malecki and Olszewski describes the changes to the internal organisation of science they believe to be accompanying its increasing interaction with technology and economy. It predicts that the institutional separation of science will diminish, as will the degree of demarcation of its pure and applied components; disciplinary boundaries too will be weakened and scientists will increasingly group themselves either around important practical problems or general techniques and types of instrumentation. In contrast, Price and Ben-David, in their contributions, regard the present institutional boundaries of science and technology as given and necessary; increased utilisation of research can therefore only be brought about by stimulating the processes of weak interaction which occur across institutional boundaries. It is interesting that both Price with his emphasis on the mobility of scientists, and Ben-David, who regards the entrepreneurial exploitation of science as crucial, stress the role of people in the transmission of ideas between institutions, and reject the once popular notion that significant 'information' can be effectively transmitted through print or other media. General sociological insight into the reasons for this theoretical transition has been provided by Burns (1969).

The shift in the institutional locale of science has focussed attention on the social role of the applied scientist. Cardwell (1957) has produced evidence suggesting that this role did not arise in response to industrial demand; it was generated by a surplus of pure scientists moving out of the German system of higher education into *general* employment opportunities in German industry. Only when these men had demonstrated the utility of their knowledge and effectively created roles for themselves as applied scientists was industrial demand stimulated.

On the whole, applied scientists today still receive the same training as pure scientists, before moving from academic institutions to become employees in industry or government. Accordingly, sociologists in the United States have analysed the

---

comparative effectiveness. For a concrete description of a 'Big' pure science laboratory, see Swatez (1970).

role of the applied scientists into two components: the expectations of his managers, and those of a colleague community assumed to be dominated by a pure science ethos. In that these two sets of expectations are contradictory the scientist in industry or government is held to experience inevitable role conflict. The final paper, by Ellis, provides a detailed review of this approach, but it belongs to another school of thought. Ellis presents empirical data suggesting that scientists in British industry are little influenced by any ethos of pure science, or indeed by any expectations located in a colleague community of scientists. He goes on to stress the fragmented and diverse nature of modern science – and of industrial science itself – and criticises sociologists who take pure science as a paradigm for scientific activity as a whole.

References

BURNS, T. (1969), 'Models, images and myths', in W. Gruber, and G. Marquis (eds.), *Factors in the Transfer of Technology*, MIT.

CARDWELL, D. S. L. (1957), *The Organization of Science in England*, Heinemann.

SWATEZ, G. M. (1970), 'The social organization of a university laboratory', *Minerva*, vol. 8, no. 1, pp. 36–58.

WEINBERG, A. M. (1970), 'Scientific teams and scientific laboratories', *Daedalus*, vol. 99, no. 4, pp. 1056–74.

8 Ignacy Malecki and Eugeniusz Olszewski

Regularities in the Development of Contemporary Science

Revised by the authors for this volume from their translation in *Organon*, vol. 13, 1965, pp. 193–212.

As a whole, scientific research is an extremely complicated process, which can be considered from many different points of view: epistemological, sociological, economic, legal and so on. Some general regularities, however, do seem to be distinguishable in the process, developmental trends determining the main directions of change within the themes and methods of research.

This paper is an attempt to describe these trends in contemporary science. Having started some time ago (and thus being objects for historical analysis), they will probably last for several more decades. They will characterize the qualitative and quantitative development of twentieth-century research, and the revolution within science, which is taking place now, before our eyes. The first trends that must be considered are those of integration.

The 'vertical' integration of science

The term 'vertical integration' is used to define the rapprochement of scientific research to socio-economic practice, and the ensuing rapprochement, within science itself, between basic, applied and development-oriented research.

Because of this type of integration, the importance of science in the economic, social and cultural development of the whole world, and particularly, of the highly developed countries, has greatly increased over the last few decades. This has found repeated expression in the declarations and programmes of parties and governments. For example, two definitions of the role of science, dating from almost identical periods, can be cited here: 'the direct productive force' whose application 'becomes a decisive factor in the powerful growth of the productive forces of society' (The Programme of the Communist Party of the Soviet

Union, 1961)[1]; the factor which 'revolutionises methods of production, changes social relationships' (The Programme of the Republican Party of the United States, 1959).[2]

Such statements can be contrasted with the attitudes taken by politicians over twenty years ago, and, in some cases, even ten years ago. Such a change was not brought about, of course, by a sudden transformation of politicians. (Although violent changes of opinion did occur, for instance in the United States after the first artificial satellite had been launched by the Soviet Union.) The underlying cause was that the feedback between scientific research and practice was gradually getting closer and closer.

Several factors determining this process can be distinguished:
1. The close connection of scientific research – both as cause and as consequence – with the rapidity of technological progress, making possible increased production and consumption of consumer goods. Motor-car production is an example of this, and also typical are the great changes occurring within the apparently very conservative food industry. This technological progress is also determined by competition in world markets, exemplified by constant advances in the construction of machines, motor-cars, ships and so on.

2. The rise in living standards, and the increased requirements for health protection and cultural services which results. Examples can be provided, on the one hand, by the struggle against epidemic diseases; on the other, by the development of television.

3. The transition from machines and technologies based upon comparatively simple laws of physics, where every success was determined by practical knowledge and the designer's intuition, to machines and technologies based upon the utilization of complex physico-chemical effects, the mastery of which requires thorough scientific study; this can be exemplified by nuclear technology or semi-conductor electronics, both of which exploit atomic processes.

4. The practical possibilities already obtained and anticipated in

1. See reference under 'The 23rd Congress of the Communist Party of the Soviet Union'.

2. Quotation from the paper of Dedijer, 1964.

the domains of technology, agriculture and medicine. These were made possible by a number of factors: nuclear investigations, the chemistry of polymers, the discovery of antibiotics, the isolation of nucleic acids and so on.

5. The tragic but indisputable fact of the immense influence exerted by armaments upon the direction and extent of scientific research in some countries.

The general laws of economic behavior, and more particularly, of the pursuit of most effective solutions, apply fully in estimating the effects of applied research. Comparative calculation is usually favorable to solutions based on new scientific results. Sometimes, however, this proves to be the case only after a certain critical point has been passed, the product obtained with the new method being initially more expensive (at times even much more expensive) than that obtained by traditional means. This was formerly the case with the production of plastics; it is true now for the production of energy in nuclear reactors. In such cases economic calculation continues to be a factor hampering any bold investment, and simultaneously forcing further research with a view to perfecting the production process.

The variety of tasks required of science by the national economy has given rise to an enormous extension of research centres. There has been a considerable increase in the number of research workers, and a still more considerable one in the costs of research both as reckoned with respect to one working post and (even more so) on the basis of absolute figures.

One of the most revealing indices for the general appreciation of the development of science, and, at the same time, of its increasing social significance, is the share of the gross national product devoted to research expenditure. Objection could be taken to the way this index is computed in particular countries; none the less it does indicate levels of expenditure on science, and changes therein, with adequate precision (Figure 1).

In general, the growth of scientific research is more rapid than that of production itself, and this tendency is showing no signs of changing. It can be predicted, therefore, that the percentage of gross national product destined for research will continue to grow. This growth will not be linear, rather it will proceed along a

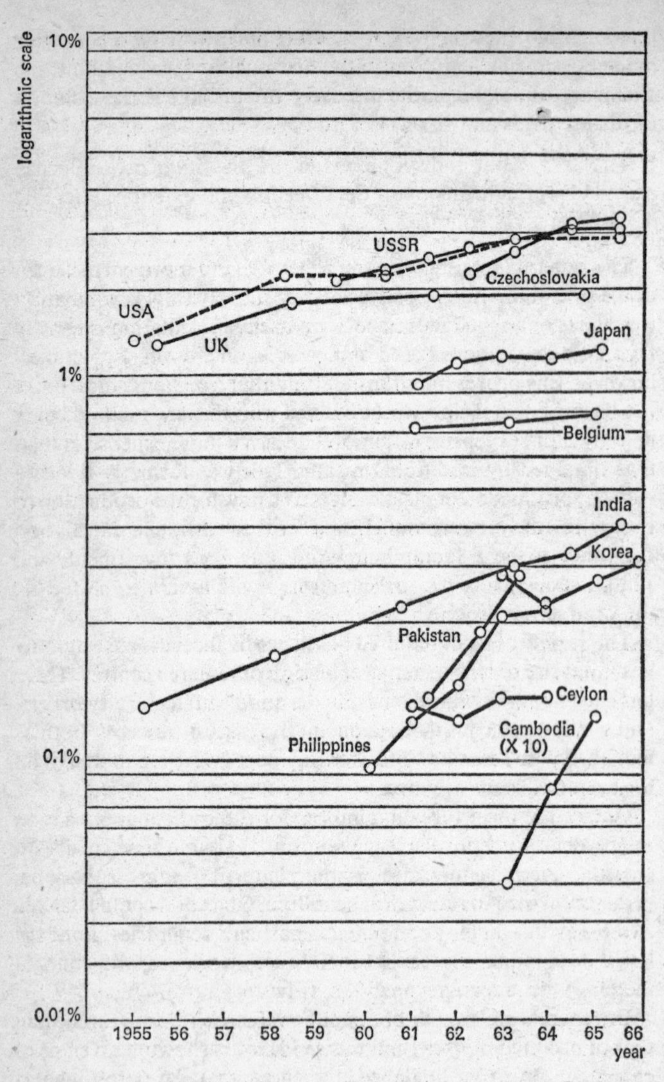

Figure 1 Research expenditure in percentage of the gross national product in some countries in 1955/1966 based on '*Document de Travail du Secrétariat de l'Unesco*', Paris 1969

logistic curve showing a point of inflection and afterwards tending towards a horizontal asymptote. (See for example de Solla Price, 1963, pp. 20–21.) Taking into account likely developments in biochemistry, the utilization of nuclear energy, the application of plastics, the exploration of outer space and so on, it seems likely that the inflection in the curve will not arrive for several years; the share of national product destined for research will probably be stabilized at, at least, 4–5 per cent.

The concentration of research, which can be seen today in the great centres of technologically leading countries, is bound up with the increasing stratification of world science to the detriment of less-developed countries.

There are, at present, enormous differences in levels of national expenditure on science; in leading countries per capita research expenditure is more than ten dollars, even more than twenty in some cases, whereas in the less-developed countries the corresponding figure is only 0·1 – 0·3 dollars. These disproportions, moreover, instead of being obliterated with the passage of time are becoming more and more marked.

This phenomenon is unfavourable not only from the human point of view, but also because it hampers feedback between research and the economic and cultural requirements of the less-developed countries. International organizations, such as UNESCO, have overcome this problem only to an insignificant degree. One factor attenuating the consequences of stratification is that the geography of the utilization of research need not correspond to the geographical distribution of research centres. In particular, international institutes may radiate influence far beyond the frontiers of the countries in which they are situated. It can be said, therefore, that appropriate popularization of research results is, at least for the present, the main way towards surmounting the disproportions in the development of science between nations, and towards completing the feedback circuit between theory and practice, in the countries now on the way to development.

This feedback brings about not only quantitative growth in research work, but also certain changes in its scope. The most characteristic change at present is the increasing significance of basic research, and more particularly the rapid development of directed basic investigations, which have not previously been common. About 7–15 per cent of general expenditure on science is now being assigned for basic research in some countries, whereas the corresponding figures twenty years ago fluctuated between 3 and 8 per cent.

The reasons for the increased importance of basic investigations reside in the very essence of the trend towards the 'vertical' integration of science. For the experience of recent years has shown that the creation of a theoretical reserve, that is, the prosecution of research not connected with the direct and present-day requirements of practice, is exceedingly worth while, since, potentially, it offers possibilities of a rapid leap towards quite new constructions and technologies. A classical example is the work on theoretical physics in the United States immediately before the decision to start atom bomb production. It is, moreover, because of the existing high level of theoretical mechanics that the hard problems of rocket and supersonic aircraft flight have been so quickly mastered during the last twenty-five years.

To increase further the links between basic and applied research is, at present, a very difficult problem, and at the same time, a very pressing one.

Deep-going specialization, which stimulates the formation of hermetic languages within particular sciences (mathematics, for example) gives rise to increasing difficulties. On the other hand, the development of collective research, exemplified by the participation of physicists and biologists in the activities of factory laboratories, paves the way for such an integration. Apparently then, a tendency to combine basic and applied research within a single scientific institution will emerge in the future, with a simultaneous deepening of the divisions between the different research teams existing in the field in question.

The growing pace of scientific research exerts a considerable integrating influence. This is especially true of areas of application and development, where the time lag between establishing the physical foundations of an invention, and its being put into practice, is steadily decreasing. Thus, photography is a case where more than a century passed before a known principle was realized in practice, and for the telephone the gap was fifty years. In this century it was fifteen years for radar, six years for the atom bomb and merely two years for the maser. In certain cases 'negative times' are encountered, where physical principles are discovered empirically in a production plant and only later find theoretical support; one example here is the discovery of the influence exerted by changes in the dislocation structure of materials on their mechanical properties.

At present, once a new phenomenon has been discovered research work begins to develop in parallel in many laboratories, resulting in a veritable avalanche of publications (this phenomenon is illustrated by Figure 2). Basic investigations prosecuted in such a way subsequently accelerate in their rate of progress, and take on a style near to that of sporting competitions, for only the first to publish given results achieve recognition for them. This rapid pace of basic research is bound up with a rapid obsolescence in its results; this is illustrated by Figure 3.

The 'horizontal' integration of science

Tendencies to 'vertical' integration are inextricably intertwined with those to 'horizontal' integration; the latter consists in the interpenetration and overlapping of traditional disciplines, and the concentration of various kinds of research around complex problems.

Specialization is often contrasted with 'horizontal' integration. These trends, however, are not altogether contradictory; they supplement each other to a great extent, for the development of teamwork methods enables various branches of knowledge to cooperate.

The 'horizontal' integration of science proceeds along many different, and frequently quite dissimilar paths; two basic forms of it can, however, be distinguished: integration around a problem and interdisciplinary integration; intermediate forms also exist.

Integration around a problem is closely connected with the 'vertical' sort. It was and is occurring in a number of practical contexts, and can be seen particularly clearly in large engineering projects and in military programmes. Only in the last quarter of a century, however, has the frequency and range of this kind of integration become so considerable as to condition in great measure the development of research, as it does at present.

An already classical example of such 'horizontal' integration is the work on radar and guided missiles dating from the period of

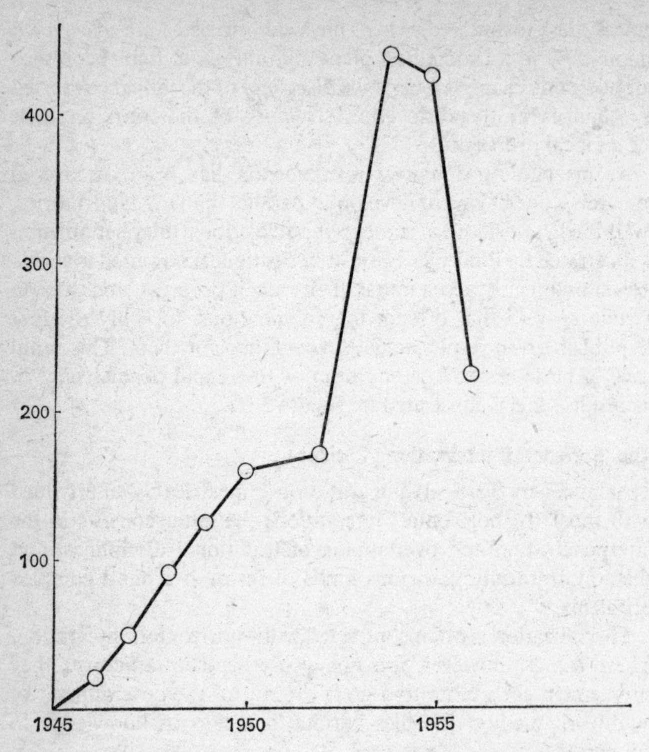

Figure 2 Numbers of articles and *communiqués* on the scope of quantum radiophysics (microwave spectroscopy and radio spectroscopy of solids and gases, radiation quantum boosters = masers) published in the American monthly *Physics Review,* in 1945 to 1956; the maximum occurring in the period of building the first gas maser. Based on an elaboration by Mrs A. Jankowska

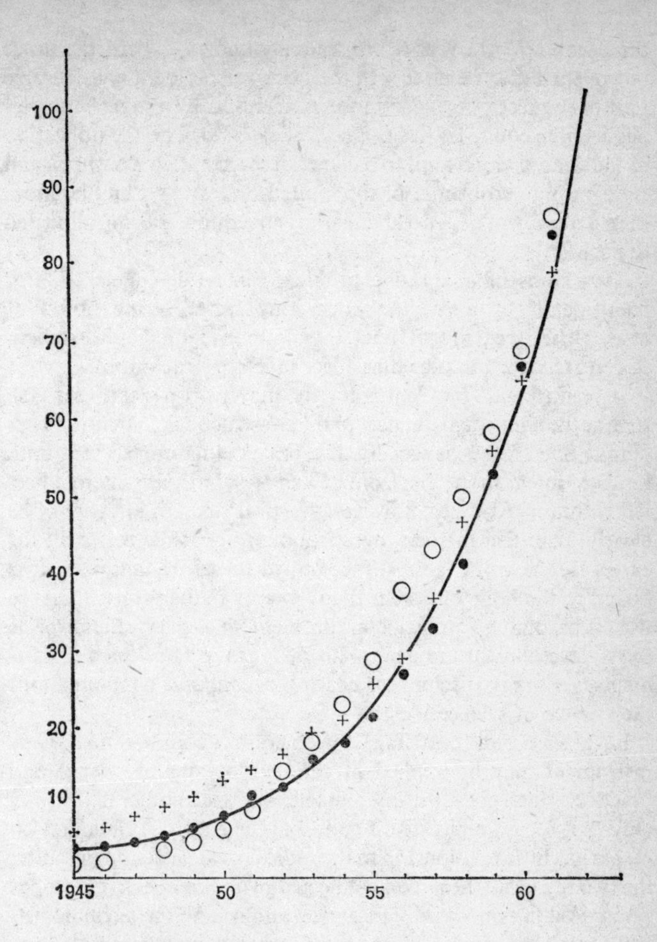

Figure 3 Percentages of citations in physics journals (April to May 1963) originating before a given year. The diagram comes from the article by S. Dedijer, as cited in footnote 2

the Second World War (as is generally known, cybernetics arose out of the latter research). In the post war period a considerable part of the research expenditure in scientifically and technologically leading countries has been devoted to work on the utilization of nuclear energy and to cosmic research, two complex and 'integrating' problems; in the United States for example, more than 8·5 per cent of general research expenditure is being allocated to each.

Two kinds of integrating problem can be distinguished. For 'centripetal' problems the integrating factor is the object of research, for 'centrifugal' ones, the common method or instrument of research (for instance the source in isotope radiation).

A 'centripetal' problem is for the most part a clearly set task such as the fight against cancer or the construction of nuclear power plants. Such a task is usually of a practical nature, so that both kinds of integration, 'horizontal' and 'vertical', are interconnected within it. At times, however, 'vertical' integration cannot be clearly identified in such investigations; in cosmic research, for example, the work is mostly cognitive in nature, although it is linked to the hope of practical advantages in the future. (Science itself is becoming a 'centripetal' problem, since the various aspects of its development are dealt with by a group of selected departments, of many disciplines, bearing a common denomination: the science of science.)

In the case of a 'centrifugal' problem, the common method or instrument may be applied in various domains of science and practice, which are sometimes linked with one another in this way only. So, for example, investigation of the effects of radiation on plants has little relationship to the radiological detection of faults; the two areas are, however, connected by the radioactive isotopes being used in both cases. The agriculturalist and the technologist, while using isotopes, are using the same language: the denotations of irradiation units, the classification of isotopes, and so on. A closer direct connection, having the character of a 'vertical' integration, occurs in this case between the agriculturalist and technologist on the one hand, and the specialist in isotope physics and isotope technology on the other. The agriculturalist, for instance, must be able to work out what type of isotope source would suit his work, and must know how to use it; the designer of isotope equipment, on the other hand, ought to be aware of the purpose his apparatus is likely to have to serve, and of the conditions in which it is likely to be used.

Integration around a problem consists more in a joint command and strategy of research than in its substantial connection, although, in fact, it presupposes a substantial link between particular sections of research. Complex investigations of this kind are carried on, as a rule, in a collective way, each of the collaborators performing some determined tasks within his own specialty while acting as a controlled and a controlling party in relation to other research workers.

In the case of such tasks, there must be, in addition to specialized panels, a co-ordinating team (or equivalent group of individuals) whose members have to possess an extensive knowledge, or at least a good background, with respect to all the questions implied by the integrating problem.

Because of the amount of expenditure which is usually necessary for conducting complex research, the initiative in taking it up most frequently resides not with scientists, but with the state or business organizations. More and more frequently, the initiation and general co-ordination of such research, simultaneously conducted in various countries, is coming under the control of international bodies. Already, today, this is the basis of the importance and authority of such organizations as the World Health Organization, the Food and Agriculture Organization, the World Meteorological Organization and others.

One specific kind of integration around a problem is the investigation of the effects of scientific and technological activities. Technology connected with science has become not only a considerable economic force, but also a force conducive to determined social changes, and a natural force as well, introducing important changes into animate and inanimate nature. In these spheres, the effects of technological activity are not always positive. It can, for example, entail an intensification of social conflicts, as when automation is insufficiently prepared with respect to employment problems, and leads to unemployment. It can also negatively affect man's natural environment. So there is a need for all-round enquiries into the effects of technological activities, both natural effects, dealt with by ecology among other disciplines, and social effects, investigated by sociologists and economists. Examples of far-flung social research of this kind in Poland are investigations of the socio-economic processes occurring around the newly arising big industrial centres (Plock, Pulawy).

Interdisciplinary integration processes are of a different character. Integration here takes place within the very process of scientific cognition, when it becomes necessary to seek methods and stocks of information, originating in various different disciplines, in order to obtain a full description and understanding of particular phenomena. The mechanism of integration may act in different ways here:

1. A new discipline may arise at the contact point of two existing ones, and draw to an equal extent from the methods and information of both; typical examples here are bio-chemistry and the history of science.

2. One of the disciplines may predominate in the 'border-strip', the contribution of the second being supplementary. So, for instance, machine designers must be familiar with problems of electronics, phoneticians with those of acoustics, and so on.

3. One discipline may use another as a research method: a typical example here is mathematics, much used in the natural sciences, and increasingly, in recent years, in the social ones too. New disciplines can arise from this combination, such as biometry and econometry. In the last decade cybernetics has started to play a role analogous to that of mathematics.

4. The fundamental assumptions of certain sciences may be combined with philosophy, and particularly epistemology. These combinations have recently increased in importance due to progress in the cognition of the micro- and macro-structure of the universe, as well as new discoveries in biochemistry and genetics. In connection with this, the interest of scientists in philosophical problems has increased, and philosophers have become aware of the need to examine the implications of the achievements of contemporary knowledge. A characteristic example of these trends is the broad discussion on this topic recently occurring in the scientific institutes of the Soviet Academy of Sciences.

With this kind of integration the interpenetration of sciences goes far deeper than with integration around problems. It is not enough here to issue orders to a neighbouring field of science; here one has to know how to use its methods oneself. Thus a team of biologists and chemists does not suffice to replace genuine biochemists, well versed in operating with methods specific to both of these sciences.

The immediate causes of interdisciplinary integration are thus the actual needs of science. In this case problems of organization and coordination arise as secondary effects. For example, the development of bio-chemical institutions was a result, and not a cause, of processes of integration occurring within science itself, on the initiative of scientists themselves.

By observing processes of 'horizontal' integration, which have been occurring more and more frequently in recent times, the following conclusions can be drawn:

1. The importance of complex problems for the general development of science constantly increases, and collective efforts are leading more and more often to the solution of these problems. The choice of such problems and of proper research strategy is thus becoming an increasingly important and responsible task. The development of research of this type leads to the rise of scientific institutes concerned entirely with such problems, replacing, in some cases, discipline-bound centres; this tendency is likely to increase in the future.

2. It is collective teamwork that is decisive for the success of complex research. One of the main problems in the contemporary organization of science is therefore to study and perfect forms of collective research. It seems clear that the international scientific organizations should undertake more extensive studies of this problem, viewing the fields of scientific activity from the standpoints of psychology, sociology and theories of organization.

3. In view of the huge resources required for complex research, international cooperation in this field is now becoming particularly important. Special protection and care must be devoted to such international scientific contacts as tend towards a common, or at least coordinated, solving of the great research problems, as is already happening, to some extent, in the spheres of nuclear and

cosmic research. Other positive examples here are the big UNESCO and ICSU research programmes, such as the Years of the Quiet Sun, or the biological programme. The research work conducted by UNESCO on the present developmental trends of science could be a valuable element in scientific-organizational activities of this type.

4. Interdisciplinary integration overthrows traditional divisions in science, and creates new branches of scientific activity. This has consequences for the organization of academic training: the basis of theoretical teaching is widened; frameworks of specialization are extended.

5. Integration around problems is likely to show the following trends in the near future:
a. Further stress on problems of basic knowledge concerning the 'cultivation' of our planet is likely; examples of such integrated 'centripetal' research already planned are: the hydrologic decade, investigations into the Earth's mantle and oceanographic research.
b. Research on basic problems will continue to develop, bent on extending the scope of science; it is here that activities in cosmic, genetic and nuclear research belong.
c. Research aimed at improving mankind's living conditions will be of increasing importance; here problems of fighting epidemic diseases, searching for new food bases (photosynthesis, the synthesis of albumens), and investigating man's ecology in contemporary conditions, come in.
d. It is probable that research intended to improve military equipment will continue for some time.

6. In the sphere of interdisciplinary integration the following tendencies seem to be developing:
a. The rise of various border disciplines between the exact and the biological sciences (biochemistry, bionics, etc.).
b. The creation of new disciplines based upon a wide utilization of mathematical and cybernetic methods (econometry, sociometry, mathematical linguistics, etc.).
c. The penetration of physics into applied disciplines, provoking a regrouping within the range of classical applied sciences;

d. An ever tighter linking of social sciences with natural and applied sciences.

The structure of the development of science and the pace of its development

[In this section of their paper, which is omitted here, the authors suggest that the increased pace of scientific development, and particularly the increased rate of incidence of Kuhnian revolutions, make it increasingly important to combat 'conservatism' in science, a phenomenon manifested, according to the authors, in over-specialisation and a dogmatic, uncritical attitude to existing theories. They regard integrating tendencies as one mechanism acting against conservatism. Ed.]

The influence of new scientific equipment

The contemporary development of new equipment for scientific work is characterized by two trends: the introduction of ever more complex and ever more expensive devices, and the increasing application of cybernetic equipment.

Investigation of intricate physical and chemical phenomena, becoming acquainted with the structure of the microcosmos on the one hand, and with the general structure of the cosmos on the other, the development of molecular biology as it approaches the discovery of the profound mechanism of the phenomenon of life, reconnaissance aimed at extending man's expansion beyond the earth – all this requires new, ever more sophisticated equipment. Its high costs, besides making research more and more dependent on the aid of the state, or on the assistance of powerful financial groups, constitute one of the elements of 'vertical' integration, i.e. of the feedback between scientific research and practice. This expenditure, often exceeding the actual possibilities of middle-sized and smaller countries, stimulates international scientific cooperation; examples here are the international nuclear research institutes (Centre Européen des Recherches Nucléaires in Geneva and the United Nuclear Research Institute in Dubna).

The development of research equipment also induces changes in the character and system of scientific work. It aggravates the specialization of scientific workers, sometimes creating an intensive division of labour. Sometimes also the costliness of equipment leads to a reversal of previously existing principles for the planning of research: the scientific worker no longer selects equipment and matches it to the research programme, instead scientific workers are chosen to work on information furnished by established equipment.

At present science is becoming transformed into one of the forces of production. Linked with this is a certain assimilation of the character of scientific work to that characteristic of a large mechanized industry; operating intricate equipment is in some of its features analogous to operating machinery. Of course the division of labour does not extend to the point where the scientific worker becomes 'a mere addition'[3] to the equipment, particularly because of his being able to rely on technical workers for assistance. Nevertheless the development of equipment does produce trends towards a narrowing of the scientific worker's qualifications, and thus towards their degradation.[4]

Another factor stimulating specialization is the constant and rapid increase in the stock of scientific information, which, first of all, finds expression in the growing avalanche of publications. Estimates show that the scientific books published every year can be counted in tens of thousands, and in the same period, at least half a million papers appear in about thirty thousand scientific periodicals. (Compare for example de Solla Price, 1963.) As a result, the part of the scientific worker's time devoted to getting acquainted with publications, i.e. not to creative, but to receptive work, grows immensely.

A considerable proportion of the scientific worker's time is also consumed by preparing material, later to be used – together with the material drawn from reading – as a foundation for creative research. The preparation of material requires creative work as well, though: the choice of testing devices and their utilization, the planning of mathematical methods for processing results, the setting of a key for archivistic investigations or statistical elaborations all involve this. A considerable part of the time in the preparation stages of research is, however, devoted to work lacking creative character: the performance of observations or experi-

3. *The Communist Manifesto* thus determined the position to which the machine was reducing the worker in a mechanized industry.

4. Arcimovitch (1964) writes about the transformation of the scientific worker into a 'skilled preparator'.

ments according to some established method, of standard calculations, the perusal of record material, the elaboration of statistical data and so on. It can be generally stated that the scientific worker spends the major part of his time on non-creative, unproductive activities.[5]

Hence the great importance that the reduction of the non-creative part in scientific work has, both for the intensification of the pace of research, and for overcoming narrow specialization and realizing integrating trends.

Measures aimed at releasing scientific workers from non-creative work were, until recently, strongly limited in nature and restricted in range. They lay in the division of work on the one hand, with parts of the preparatory tasks being transferred to attendants and laboratory hands, and in the rationalization of library and bibliographical information, the publication of abstracts, the acceleration of calculations by the introduction of logarithms (and of the slide rule and arithmometer), and the application of standard measuring devices, on the other.

However, to an increasing extent in recent years, new revolutionary technical devices are being produced, which can, and doubtless will, introduce essential and qualitative changes into scientific work. These cybernetic devices are, today, already able to take over and multiply the functions of human memory, gathering and storing huge amounts of information in ways that facilitate swift and convenient data retrieval. They are also able to transform information by applying mathematical and logical operations. The use of cybernetic equipment in accelerating calculations, translating scientific texts, seeking optimal solutions and performing medical diagnostics is already well known. Attempts are also being made to employ it in historical research. (See for example Dobrov, 1965.)

Automatic measuring devices are becoming more widespread, which can convey measurements directly to cybernetic units for storage or transformation. Similarly visual observations can be made, without human participation, for instance by the equipment of artificial satellites; results can be stored on film, and

5. For historical science it has been estimated that the time of creative work barely amounts to ten per cent of a scientific worker's time destined for research. Compare the article of Dobrov (1965).

eventually, in a suitable transformed state, be preserved in the memories of cybernetic devices. Particularly important here is the automation of large and expensive equipment, which releases scientific workers from tedious tasks involved in its operation, and also makes possible the acquisition of data from temporarily inaccessible areas such as the surface of the moon.

To an increasing extent with development of cybernetic devices, the non-creative parts of scientific work, and more generally of intellectual work as a whole, are being transferred, leaving the human mind free to concentrate on creative tasks. And since the memory of cybernetic equipment is generally larger and more efficient than its human equivalent, there will eventually be no need to charge man's mind with a ballast of information better conserved and subjected to elementary transformations within cybernetic devices.

Of course, at present, the range of operation of cybernetic equipment is still far from its theoretical limits, and many a real technical solution is economically unrealistic. Such devices are, however, setting the direction of development for subsequent decades by leading towards a reduction in the percentage of time devoted by scientific and intellectual workers to non-creative work, just as automation is releasing the manual worker from his role of being a 'mere addition' to machinery.

Prospects of this kind ought already to be having effects on methods of higher training, and in particular on the methods of training scientific workers. Once, when asked what was the speed of sound, Einstein replied that he was not in the habit of charging his memory with information readily obtained from encyclopaedias. Modern cybernetic equipment and its anticipated development, make it easier to eliminate the ballast of rules and symbols from training programs and examination requirements. The time and talent thus released can then be devoted to developing creative mental factors which can be replaced by cybernetic devices only to a very limited extent.[6] This will make it possible, at the same time,

6. A separate discussion is required by the question of to what extent the stock of information recorded in the human memory is indispensable to the creative work of the mind, since it is obvious that the shifting of the whole stock of information to the cybernetic memory will never be possible. The question of the mind's adaptability to the new working conditions will require, however, suitably thorough investigations.

to overcome trends towards narrow specialization, and gradually to extend the scientific horizons of young research workers.

Thus science, which is inseparably linked with technology in our era, is supplying by itself the means and measures for overcoming the trends to an exaggerated specialization which threaten its development, and is creating conditions that favour integration, and hence the fight against dogmatism also. It is the task of science policy, particularly as applied to the training of scientific personnel,[7] to ensure rapid exploitation of these possibilities. In this problem among others, we recognize the importance of a new scientific discipline – the science of scientific activity.

7. Some problems connected with this are dealt with in a more extensive way in the paper by Olszewski (1964).

References

ARCIMOVITCH, L. A. (1964), 'Some regularities in the development of physics', in *Dialectical Materialism and Contemporary Science*, Moscow.

DEDIJER, S. (1964), 'International comparisons of science', *New Scientist*, no. 379.

DOBROV, G. M. (1965), 'Research in the history of technology and cybernetics (Badania historyczne techniczne a cybernetyka)' in *Quarterly Journal of the History of Science and Technology* (*Kwartalmik Historii Nauki i Techniki*), no. 1.

OLSZEWSKI, E. (1964), 'The structure of the development of science and the existing system of training scientific workers (Structura rozwoju nauki a istniejacy system ksztalcenia piracownikow nauki)', in *The Life of Higher Schools* (*Zycie Szkoly Wyzszej*), no. 12.

SOLLA PRICE, D. J. DE (1963), *Little Science, Big Science*, Columbia University Press.

The 22nd Congress of the Communist Party of the Soviet Union (1961), *XXII Zjazd Komunistyczneg Partii Zwiazku Radzieckiego*, Warsaw, pp. 573 and 630.

9 Derek J. de S. Price

Science and Technology: Distinctions and Interrelationships

D. J. de S. Price, 'The structures of publication in science and technology', from W. Gruber and G. Marquis (eds.), *Factors in the Transfer of Technology*, MIT Press, 1969.

The evidence I should like to bring to bear upon our problem of transfer is that which has arisen in a series of researches whose aim has been simply to try to understand the nature of scientific work, historically and in the present. It is perhaps worth underlining the point that our work derives from scholarly curiosity rather than from the more usual objective in such investigations: an attempt to do some good, to cure some evil, or to find out how to organize something efficiently. I am therefore in a strong position to present evidence which connects reasonably well with what little we know about the history of science and technology, the sociology of science, documentation theory, or the like, but in a correspondingly weak position if one wants anything like a specific answer for practical means of organizing technology transfer.

Dynamic structure of science

The body of research that I believe is most relevant derives from the quantitative investigations of all the things about science to which numbers can be attached. There has been a long tradition of such work,[1] counting papers and measuring manpower, expenditure of money, and several other things, but most recently a better set of quantitative data than ever before has become available. As a by-product of the commercial production of Garfield's *Science Citation Index*, we have available a corpus of several million papers, citations, and authors which are very readily susceptible to investigation. This very large population gives one a statistical certainty that was quite unreachable with the older and cruder head-counts, and we are now in the position

1. See Price (1963) for a bibliography.

of being able to link together and crosscheck several different sources of data and types of measurement.

Crucial to the whole investigation is that we can first derive reasonable proofs for the validity of these head-counts as a legitimate measure of scientific inputs and outputs. We have been able to show that in spite of the obvious lack of any control over the quality of the heads one counts, the same statistical laws are followed which one could get if all papers or persons were weighted by their magnitudes on any reasonable scale. The detailed results might be very different for any particular individual, but the distribution of the whole population would be unaffected.

Working from the population of papers, then, treating each as a sort of atom of knowledge, we have been able to derive something of a model to show how new papers are related to old ones. Since each paper carries an average of about a dozen citations back to past literature, these citations may be analyzed to find their pattern. Such analysis shows immediately that there exist two separate modes of citation (Price 1965a). The first, accounting for about half of all references back, may be called *archival*. It represents a raiding of the archive, almost completely independent of the age of the older paper being cited, and without structure. The absence of structure occurs as a random and patternless set of connections between the new papers and the entire body of old papers in that particular field. Some papers of course are more popular, more cited than others, but highly connected recent papers will probably contain quite different selections from the archive.

The second type of citation may be called *research front*. New papers use the other half of their references to connect back to the relatively small number of highly interconnected recent papers. In a particular field each recent paper is connected to all its neighbors by many lines of citation. A convenient image of the pattern is to be found in knitting. Each stitch is strongly attached to the previous row and to its neighbors. To extend the analogy, sometimes a stitch is dropped and the knitted strip then separates into different rows, each of them a new subfield descended from the first.

As the first fruits of this model we may derive some explanation of how it is that science has had, for several centuries, an uncannily constant rate of exponential growth of the literature. The number of scientific papers in all fields and for all countries, with only tiny perturbations for world wars, has been doubling every ten to fifteen years – depending on just what is measured or how it is measured. In our theory, each scientific paper ever published, lumping together good ones and bad ones, produces on the average one citation per year. Since it takes about twelve such citations to make a new paper we have a growth rate of about eight percent per annum, a doubling about every decade.

There are, therefore, grounds for believing that science grows as it does, so much faster than mere people, because old knowledge breeds new. Furthermore, the key process seems to be that in which very recent knowledge breeds new so much faster than it does when it later becomes packed down into the archive. Science grows very regularly, in a very structured way, and from its epidermis rather than from its body. In terms of our present problem this model may be taken as giving a mechanism for the 'transfer' from old science to new science.

The model may be further analyzed to give a little more of the nature of this, the simplest transfer that we have to consider. It has already been admitted that some papers are better than others, and equally truly some researchers are more powerful than others. An investigation of the distribution of quality in men and in papers shows immediately that there exists greater inequality than in any country's distribution of economic wealth and a bigger rat race than in any other competitive activity of mankind. The gap between a Nobel prize winner and a plain person is rather larger than that between an Olympic gold medalist and an ordinary mortal. Roughly speaking, if you have published n papers in your life, the chance of your reaching $2n$ happens to be about 1 in 4. As an inevitable consequence of a completely lopsided skewed distribution of this sort, it follows that a small number of the men who write scientific papers happen to be responsible in the aggregate for half the production and a good deal more than half of the value if one weights each paper in any reasonable way. This small number, a hard core, may be estimated at about the square root of the total number of writing scientists, and together they constitute the heart of what has become known as the Invisible College of all the good people who really count in that neck of the scientific woods.

Now the important thing for transfer studies is not so much that an Invisible College exists, but rather that with such a high rate of mortality, the large majority of people get lost or excreted while trying to make the grade to get into the core. This process is what we call Graduate Student Education. The mortality is so high not only because the work is difficult, but no matter how much the old science is packed down, the poor student has to get through the archive of centuries and then run much faster than the rate of progress of the research front in order to emerge at the growing edge of science. When he emerges with that particular training, that particular mode of approach, those little bits of research front material passed through, he will then (with luck) be able to make his new contribution at just that particular area of the research front. If he then lives with it for a while, always keeping up with his cohorts at this place on the wide front, he will participate in the knitting of new knowledge to the old which he and his friends have laid down.

Perhaps one should emphasize that this is the normal growth pattern of scientific research and most published work occurs thus. What does not usually happen is the *ab initio* growth of new knowledge coming from almost nowhere. Only in that convenient mythology of science that historians spend their lives trying to dispel, does it appear that the mountain-peak contributions of Newton, Copernicus, and Galileo arose from the native genius of isolated minds. Innovation in the sort of knowledge that is published in scientific papers arises when new facts, new experiments, new theories are added to the immediately preceding old ones in a very structured way. Transfer, from outside the knowledge of men is a rare thing, though it may be very important when it happens. I hazard a guess, however, that even when some outside force, like the invention of a cyclotron, makes knowledge growth burst out in new ways, it does so only in retrospect as we attribute a magnified power to the single great event and forget the less dramatic build-up of the host of related researches clustered around the clear main line.

The place of publications in science and technology

Armed with this sort of model for science, we may next turn our attention to technology. (Immediately there arises the hoary problem of defining science and technology. The usual course is to pick a definition from thin air – and then never use it.) With this model we have an interestingly new approach (Price 1965b). We have been talking about science in a way such that there is an implicit definition for it. Science, to fit this model, must consist of the scientific papers that are being cited, counted, and otherwise manipulated in such studies. I therefore propose, as a formal definition, to take as science that which is published in scientific papers. We may define these in turn, either crudely as articles in journals in the *World List of Scientific Periodicals*, or more artfully by making use of the knitted research-front structure. We may suppose quite reasonably (and the evidence bears it out) that the structure is what makes science different from the nonscientific scholarship that is published in other periodicals that are probably not in the *World List*. Along the same line of definition we shall define a scientist as a man who sometime in his life has helped in the writing of such a paper. If you complain that these definitions are too loose, I can tighten them as much as you like from the same theory. If we know the number of 'scientists' by my definition we can compute, let us say, the number of men who have published at least one paper a year over the last three years. If we know the total number of papers we may fairly easily compute the number that will ever be referred to in a subsequent paper by anybody except the original author.

It should be said at this point that this definition is useful because it can be used to tie together most of the known sets of statistics about pure science, and because it agrees very elegantly with the important sociological analysis of scientific private property, multiple discovery, and priority disputes. This analysis, due largely to the work of Merton (1962, pp. 447–85), is now a cornerstone in the historiography of science and fits well with the best integrated views of what happens in a scientific society. Because there is after all only one scientific world to discover, it is terribly important to lay claim to the discovery by publication. Beethoven's symphonies would never have existed without the man, but Planck's constant might just as well have been discovered by Professor Joe Bloggs. Indeed, scientific papers turn out to be such a good indication of science simply because the high motivation to publish exists because publishing is the only external sign that work has been done that can enter into the research front, make new knowledge, and hopefully (though more rarely) give the originator the immortality of his work in a perpetual archive.

With science defined in this way, what is to be the position of technology? Certainly not more than a part of technology could possibly be subsumed under such a definition. In the fields of electronics and chemical engineering there is, one can admit, considerable publication, but in many other well-known areas of technology there is no equivalent of a scientific paper. There are patents, of course, and these even carry citations back to previous patents. And in spite of the lack of technological papers that look like scientific papers there is an enormous technical literature – in fact, many more journals exist here than in the fields of science as it has been defined.

The difference between science and technology in this new sense is pointed to most clearly by the sociology of publication. The scientist, it was remarked, is heavily motivated to publish – this is the key to all the inner springs of his drive to do science. In technology it is otherwise; the tradition, crudely speaking, is to conceal in order to have a new product or a process before others. It also happens that the scientist hears from his Invisible College about the new work on which he is going to build, and he therefore does not need to read the published journals very much, for by then they are old hat. However, in the other camp, the technologist is most eager, while revealing as little as possible himself, to pick up anything and everything that may be dropped his way by others. One may put the whole thing in an aphorism I have used before; the scientist wants to write but not read, the technologist wants to read but not write.

If technology is, so to speak, *papyrophobic*, we cannot use the measure of papers as we could with *papyrocentric* science to diagnose the features of a model for it. When one makes the attempt to analyze citations in most technical journals and patents, one finds no well-marked structure. Papers and patents at the best form short chains, each item taking off from a preceding one usually by the same author or company. It seems pretty clear that even though there exists a large mass of technical journals, the writing does not have the same function that it does for science. It seems to exist for a newspaper-like current awareness function, for boasting and heroics, and probably, above all, as a suitable burden to carry the principal content of advertisements which together with catalogues of products are the main repositories of the state of the art for each technology.

A point which tells particularly heavily for me, as a historian of technology, is this opposite polarity of science and technology in their attitudes to literature, which is precisely what makes it terribly difficult to write the history of technology in the same manner that one writes the history of science. The content of science is already embodied in papers, whereas that of technology first has to force itself into written form, and this leads to exactly that type of antiquarian writing which so often bogs us down. It is the preliminary stage to the writing of a history of technology, but it is not the history itself.

The dynamic structure of technology

Since it seems obvious that technology is related in some way to science, and that it grows at much the same rate by any reasonable measure, it must follow (or lead) the cumulating structure of science. Now, the naive picture of technology as applied science simply will not fit all the facts. Inventions do not hang like fruits on a scientific tree. In those parts of the history of technology where one feels some confidence, it is quite apparent that most technological advances derive immediately from those that precede them. It may seem to happen from time to time that certain advances, particularly the spectacular and anomalous mountain peaks, as with science, derive from the injection of new science. In the main, however, old technology breeds new in just the same way as the scientific process already described. Because of this I now hypothesize that though it cannot be diagnosed by papers, technology has a structure that is formally identical with that of science. We shall define technology as that research where the main product is not a paper, but instead a machine, a drug, a product, or a process of some sort. In the same way as before there will be a highly competitive rat race, a similar exponential rate of growth, an archive of past technology, and a research front of the current state of the art.

One must enter the caveat that by choosing definitions that can do this work for me and by erecting such models, I have had to use words in a precise but not entirely ordinary way. Thus if a man is paid by a pharmaceutical company to develop a new drug but finds himself publishing a paper about the new chemical knowledge gained, he has been guilty of doing science on company time; but if a pure mathematician comes up with a new way of designing switchboards, it is technology, however pure be his mind.

Parallel structures of science and technology

At any rate, with these definitions and a hypothesis we now have conjured up a pair of similar and parallel systems for science and technology. What makes them run in step, or in Toynbee's (1962, vol. 1, p. 3) apt image behave like a pair of dancers to the same music, though it is imperceptible who is leading and who following? The Marxist answer would be to suppose that the music is the wishes and needs of society that make a certain type of science and a certain type of technology appear at any particular time. This, however, is quite inconsistent with the previous finding that it is the old knowledge rather than the wishes of society that is breeding new knowledge. One can, I suppose, create technology to order, just by wishing it. But ordinarily one is severely constrained by the old technology's having or not having the capacity to breed a particular desired thing. However much one is willing to pay for a cancer cure, one cannot develop it unless the state of the art, and perhaps the state of science, seems able to sustain it.

Interaction of science and technology

Rather than supposing that an outside force affects both dancers, it seems more reasonable to think that their action upon each other keeps them in step. Since we know so many cases in which science has passed into technology, and so many other cases in which the technology has made new science possible, it follows that we must have a complete interaction. The well-known lag between scientific and technological advance would seem to indicate that the dancers hold each other at arm's length instead of dancing cheek to cheek. To use the more precise language of the physicist, the relation between science and technology seems to be a weak rather than a strong interaction.

The mechanism for such a weak interaction is not hard to find.

When a technologist is educated in the research front of the state of the art, he is necessarily subjected to some training in the ambient state of the science of his time. Similarly in his education, a scientist is taught the ambient state of technology. It follows then that men on the research fronts of science and technology will be able to use each other's ambient knowledge. It seems too that this will generally be the ambient knowledge that is on the average about one generation of students old – perhaps ten years.

We have shown already that 'transfer' from old science to new goes through the knitting process of cumulation. A similar process will therefore exist for the 'transfer' from old technology to new. It will occur to a certain extent as part of the archive through which each generation of students grows up. More important, it will occur through the knitting-together process in which the man making the new technology has just had contact, often through the Invisible College, with those who have just been laying down the immediately prior technology. Last, there is the most important type of 'transfer' that occurs from science to technology and vice versa. Again the medium of transmission is the person, and the method is that of the formal or informal education which a man gets in the current state of the art in science or in technology.

Here one must interject a remark of considerable importance to science policy: that the cumulations of science and of technology habitually interact, though weakly, in such a way that they may be said to exist in symbiosis, feeding from each other. There seems to be a body of pathological case histories in science and technology indicating that whenever society has tried to force science or technology away from such symbiosis, the results have been disastrous for both. Science without the byplay of technology becomes sterile, and in the several cases in which a materialistic society decided it would pay for the technology which gave economic gain, and neglect the science which was just a chess game, the technology became moribund. If only there were stong interactions between science and technology one could point eagerly to the ultimate clear utility of pure science. With a weak interaction at present one can only have faith, though there has always been strong temptation to lie through one's teeth and claim direct benefit for pure science.

The mechanism which has been suggested to explain the interaction of science and technology has the important feature of noting that the lag between them is due to the way in which people are brought to the research front either in the science or in the technology, carrying with them the ambient technology or science of about a decade earlier. It has often been suggested by economists that this time lag was once very much greater than it is now, and that it has been decreasing markedly. As a historian of technology I cannot agree with this, and I feel that some error has been introduced by looking principally at those traumatic peaks of spectacular change, like penicillin and transistors, which are not very representative of normal interchange between science and technology. We can all agree, however, that there is a considerable lag in the weak interaction between science and technology, and that this is in contrast to the rapidity with which new technology arises directly by strong interaction from old technology and to the similar process by which old science generates new. The research front is basically contained in people, and the archive is somewhat distant from them.

It has often been supposed that some sort of communication difficulty holds back the optimum growth of technology. In terms of our new model this can be seen as something of a red herring. The papyrophobic character of technology appears to be a deep and long-standing tradition. Even in the Soviet Union, where industrial development is public property, the literature in technology seems to be vastly different from that in the scientific tradition. In the United States the fashion has been to try to force technology into the same literature cumulation as science by the device of the research report. Basically this represents a back-door method of publication, the output being forced as a fiscal device consequent upon the spending of public money, and the mechanism of report publication being used because there are no journals to take such material; not being commercial, it may not be published as a book. Evidently nobody really wants the material enough to buy it, but it has to be published just the same. There is simply no need for literature cumulation at the technological research front in the sense in which we use the term.

The effect of this unwanted research report literature in technology is probably to clog the information channels, and it should not, I feel, be encouraged. Publication should always be held as a privilege consequent upon having found something that somebody else would like to use. It is not a duty consequent upon the spending of public money. In any case, what communication difficulty there is seems due to the fact that though the scientists want to write and the technologists want to read, the scientists are writing for their colleagues in science, or sometimes for their imaginary archive; they are simply not writing the sort of material that the technologists want to read. This frustrates the technologists and makes them believe that somewhere in this pile of material, if only they could find it, there is the very valuable material they are looking for to make new products.

Support of science and technology

Next let us consider the role of the spending public and private moneys in the purchase of research and the promotion of transfer. With science, it seems clear from the complete internationality of scientific knowledge and from the absolute steadiness of the growth curves of science that the breeding of new knowledge by old is hardly affected by the expenditures of any individual nation. If the United States put an embargo tomorrow on all pure research the world's growth might be abated slightly, but the growth rate would still continue at so near the present level that in very few years the United States would have no scientists and no teachers, and therefore no engineers who could monitor the new world position of the research front. Its technology would therefore dry up in the normal time lag – perhaps a decade.

Figure 1 provides an illustration of this theory from the data of Milton and Johnson showing the trends in the sources of support for a fair sample of scientific papers published in the United States 1920–1960. The growth in numbers of papers during this interval was completely regular apart from the local dip during the second World War. The sponsorship, however, changes very suddenly in its trends; at the onset of the massive grant program at the end of the war there begins a swing that is just now tailing off because it cannot go much further. The government instead of the universities picked up the tab, but there was virtually no change in the normal rate of growth of scientific papers. Of course, one might argue that if the government had not picked up the tab the resources of the universities would have been soon stretched to

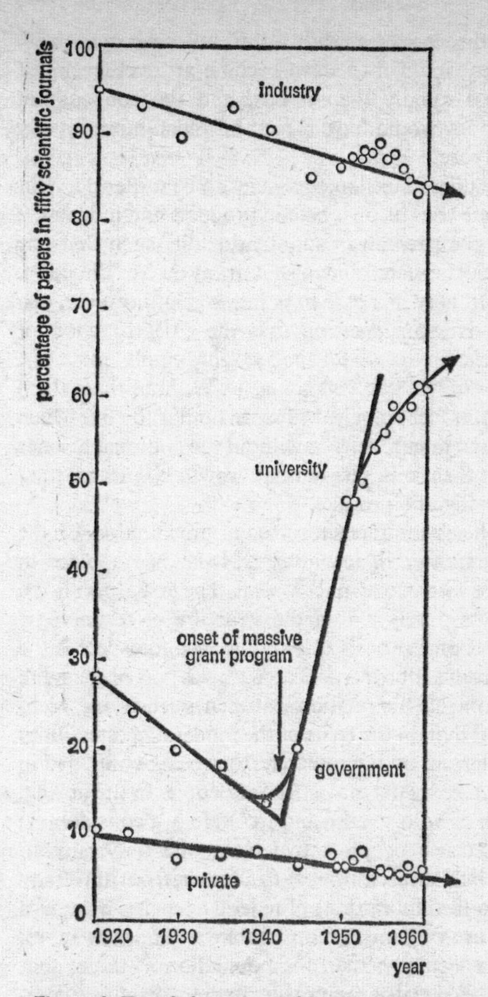

Figure 1 Trends in sponsorship of published papers in fifty large scientific journals (circa 1,000 papers in 1920; circa 10,000 in 1960).
Source: Helen S. Milton and Ellis A. Johnson, Technical Paper ORO-TX-42, Operations Research Office, John Hopkins University, September 1961, and private communication

their limits, and the normal growth would not have continued. The United States would then have become an overdeveloped country much less rapidly than it has, and the concomitant growth of technology would have sloped off much more rapidly than has been the case.

The role of spending in technology seems to be rather different because each country has its own base in products and industries; there is not quite the strength of supranationality such that one can buy technological research and push it in any desired direction. The space program is such a case in point. Again, however, one might make the daring suggestion that the chief function of NASA is that it picks up the tab for the necessary employment of a very large number of research technologists. Without them there would be a cut-off in technological growth similar to that which would have occurred in pure science without the National Science Foundation. With them one gets normal growth by means other than the direct buying of a product.

Seen in this light as the repository for a large section of the research front manpower in technology, NASA has a place in technology transfer with two distinct aspects. The first aspect is the transfer that occurs purely within the structure of technology, creating new results on the basis of old. The second aspect is the transfer that occurs in a more complicated way as part of the weak interaction that controls the relation between science and technology. It is crucial that on the basis of the model suggested, both types of transfer depend on the research front being contained in people and transferred by the movement of people. In the normal patterns of growth of both science and technology a considerable transfer occurs because people move from one university to another, from one industrial laboratory to another, from university to industry, and to the laboratories of federal agencies. Scientists and technologists are very much more mobile and migrate more, physically and intellectually, than the population at large; that they do so is probably due in large measure to the enormously competitive rat race of excellence which governs their careers. In a laboratory or institution that is growing only at the general approximately 7 per cent per annum growth rate of the overall structure of science and technology, it is usual to find that the increase comes about by something like a 17 per cent inflow of new staff and a 10 per cent outflow of old ones each year; it is this outflow that causes much of the human movement that creates transfer. With an agency or laboratory that is growing much more rapidly than this, very much less of the inflow is excreted again, and so the linkage is only at the beginning. With respect to NASA, if it does not fall on bad times that lead to massive departures from its research staff, there will be good linkage with the universities and schools that train its scientists and engineers, but there will be poor feedback of people to other industries. Perhaps one of the ways of engineering more massive transfer from NASA and defense technologies to private industry would be to encourage as far as practicable a larger continuous flow of people from government payrolls back to the training grounds, and most vitally, back to private industry.

A very similar suggestion has recently been made by Kapitza (1966) as a remedy for the falling scientific and technological productivity in Russia. He notes that there is far too much stability in the scientific establishment and that it would be generally helpful to fire people rather extensively so as to create more openings at the lower end. This means of course, not any decrease in the number of scientists and technologists employed, but a large increase in mobility and consequently a considerable increase in the communication that is effected by people acting as containers of the research front tradition and state of the art in science and technology. It seems indeed very likely that both in the US and in the USSR such a policy might come into effect whether one likes it or not for this particular purpose. Now both of the major scientific countries of the world are very rapidly filling with science and technology to the point where it is becoming increasingly difficult to muster more high-talent manpower and a bigger fraction of the GNP to research ends. Both countries are in this sense already overdeveloped, and the growth rates of science and technology in both countries seem to be dropping more and more rapidly behind the world total of growth. I would estimate that both countries are now about 10 per cent below the levels that they would have reached with uninterrupted exponential growth. As this gap widens between the leading countries and the rest of the world which must needs slowly overtake them, there will necessarily be widespread changes by those leading countries in the deployment of their fixed stock of precious resources for research. This should lead to a considerable movement of personnel from one area to another, and if this movement can be engineered with any sagacity it might well increase the transfer in science and technology to the point where the increase in pay-off compensates for the decreased possibility of absolute growth in man-power and money.

References

KAPITZA, P. L. (1966), *Theory, Experiment and Practice*, Moscow: Znanie, Series 9, part 5.

MERTON, R. K. (1962), 'Priorities in scientific discovery: a chapter in the sociology of science', in B. Barber and W. Hirsch (eds.), *The Sociology of Science*, Free Press.

PRICE, D. J. de S. (1963), *Little Science, Big Science*, Columbia University Press.

PRICE, D. J. de S. (1965a), 'Networks of scientific papers', *Science*, vol. 49, pp. 510–15.

PRICE, D. J. de S. (1965b), 'Is technology historically independent of science? A study in statistical historiography', *Technology and Culture*, vol. 6, no. 4, pp. 553–68.

Science Citation Index, published quarterly and annually by the Institute for Scientific Information, Philadelphia, Pa.

TOYNBEE, A. J. (1962), *Introduction: The Geneses of Civilizations* (*A Study of History*, 12 vols.), Oxford University Press.

10 J. Ben-David

Scientific Entrepreneurship and the Utilization of Research

Excerpt from J. Ben-David, *Fundamental Research and the Universities*, OECD, 1968, pp. 56–61.

The relationship between scientific research and technological growth has been viewed in two ways. Everyone agrees that science may benefit technology; but according to one view such benefits are a matter of random chance and expenditure on research is therefore considered as consumption and not as investment. The other, currently more favourable, view regards the existence of a systematic relationship between scientific and technological growth as proved, and expenditure on science is therefore considered an investment. It will be suggested here that the existence or absence of a systematic positive relationship between scientific and technological growth is not something given and inherent in the nature of the case, but a state determined by entrepreneurial activity and certain other conditions still to be identified.

The assumption of the economic importance of research derives from calculations which show that if one wants to account for the output of different national economies in terms of all the known inputs from the traditional factors of production, there still remains a large unaccounted-for residue. (See Denison, 1964.) It is assumed that this represents the contribution from the growth of useful knowledge. This knowledge, however, may consist of technological or organizational improvements, or the quality and extent of education. There is a whole host of examples bearing out this contention. The industrial revolution in England, or the rapid industrial development of Belgium and the United States in the last century and of Japan in this century, had not been preceded by any noticeable upsurge in scientific research, fundamental or applied. Japanese industry has done very well, in spite of the adverse balance of Japan's trade in technological know-how. A recent study of patents in the United States by Schmookler (1966, especially pp. 165–78) has produced very strong and systematic evidence that technological innovation in four important industries, railroads, agriculture, petroleum refining and paper making, occurred in response to economic demand and *not* as a spin-off of fundamental research.

Even in technologies which are directly and undoubtedly derived from scientific research, the practical exploitation of the results does not necessarily take place in the same country where the fundamental work was performed. A good example is provided by the applications of nuclear energy. This outstanding science-based technology of recent times derives directly from fundamental research in physics conducted between about 1900 and 1940, mainly in Western Europe. The practical applications of this research however took place to a much larger extent in the United States and in the USSR. Of the Western European countries which played an important role in the fundamental work, only Britain then proceeded to do important applied work.

All this does not contradict the econometric conclusions about the residual factor in economic growth, or the long-range importance of scientific research and dissemination of scientific knowledge in economic growth. It does however suggest that at least until recently there has been a very poor relationship between the place and the time of the production of basic scientific ideas on the one hand, and the reaping of technological and economic benefits from it on the other. In fact, one possible way of interpreting the evidence would be that, from the point of view of the expectation of economic returns, fundamental research is a bad and unjustifiable investment for any given country since the possibility that its results can be turned into useful technological applications is unpredictable, and it is particularly difficult to predict whether the applications will benefit the economy of the same country or those of its competitors. (Had international trade been as free as scientific communication is, this would, of course, not matter.)

From such an interpretation it would follow that, provided economic gain were the only consideration, the best policy for each country would be to reduce its expenditure on basic research to the minimum necessary for the training of people capable of 'parasitizing' the results of research done elsewhere. In practice this would imply considerable expenditure on scientific education at the secondary and higher level, with just enough overhead for research to maintain the efficiency of the latter.

There is therefore no reason to assume the existence of a generally positive relationship between fundamental research and industrial growth. It seems, however, that there *has* been such a direct relationship in some countries, such as in Germany during the second half of the nineteenth century before the effects of the rigidity of its science organization had become felt, and in the United States today.[1] The findings of Schmookler, as well as the impression from the historical cases, show that these cases have not been due to the effective guidance of fundamental work by criteria of applicability, but have been the result of effective entrepreneurship (and presumably, such related conditions as the size of the market).

The sociological aspect

We shall turn now to the sociological aspect of the problem, and try to show that the entrepreneurial approach is also more consistent with the way research is actually performed and turned to practical uses than the common-sense approach. At least so far, scientists have found it more congenial to be guided by the logically inherent problems and methodologically determined potentialities of a 'paradigm' common to a group of scientists (a scientific community), than by considerations of practical use. They have become stale and sterile when subjected in their work to extraneous considerations over a prolonged period of time. They worry about the solution of logical problems, and their scale of priorities is, and has to be (if science is to be good), determined by the intrinsic intellectual qualities of the problem and not by its practical usefulness. (See Hagstrom, 1965, pp. 9–58.) This is not true of technologists, for whom practical efficiency matters, and who for that reason often have the same aims as businessmen. There is, in fact, an uninterrupted social continuity between technologists on the one hand and businessmen on the other, as both tend to merge into management.

1. For Germany's research in industry see Cardwell (1957, pp. 675–6) and Haber (1958, pp. 128–36).

Accordingly, scientists in a given field usually have reliable estimates of the quality and prospects of the work done in their respective fields but are not particularly well informed about the prospects of technological applications of the work in their field, and are not concerned with its economic potentialities. The converse is true of technologists and businessmen. It is unlikely, therefore, that the actual relationship between fundamental science and its eventual applications in productive technology should be of the kind that one tends to imagine on the basis of logical reconstructions after the event. These reconstructions, biased by the language habit which speaks of 'application', make the technological invention appear as if 'implied' in the fundamental discovery. Once the latter is made, it is only a matter of correct 'application' to exploit these 'implications' for practical purposes.

In fact, nothing is 'implied' in a discovery beyond the questions answered by it, and those to which it is related by the traditions and mental habits of the people who are its prime consumers. The circle or 'community' of these 'prime consumers' is very limited. According to the estimates of Price the scientific community in any field does not contain more than a few hundred active members all over the world (this in addition to such passive users of science as student teachers, professionals and occasional amateurs). The reason for this is that it is impossible to follow the work of a more extended group. Hence efficiency requires the isolation and the limitation of the span of attention of scientific communities from each other.[2]

It would be even more senseless to try and link these small scientific communities to what can be called a 'technological community'. Very rarely is such a community directly and overwhelmingly linked to a field of enquiry practiced by a scientific community. Engineers, physicians or other professionals making use of science for practical purposes have a range of problems, theories, methods and techniques which consist of the use of scientific elements in ways peculiar to that community, traditions which are based only on experience, and business

2. On the size and dynamics of the 'invisible' colleges, see Price (1963, pp. 62–91).

practices concerning the marketing and the use of their products and services.[3]

The separation of the scientific from the technological and business communities explains the absence of direct links between fundamental and applied research. Of course it can, and has been, argued that this separation is not an optimal state of affairs, and that by redrawing the boundaries of these communities, a better background would be created for the practical exploitation of research.

Furthermore, it can be pointed out, with a great deal of justification, that the difficulty of establishing a link between fundamental and applied work is not symmetrical. The link from economic and technological problems to fundamental research is more predictable than from fundamental research to economically useful technological innovation. As the range of science is much more limited than that of productive technology, it is easier to see which branch of science is relevant for the solution of any technological problem. Of course, from the point of view of any given branch of science the problem may be trivial in the context of existing paradigms. Still, a technologist in search of a solution to a basic problem will more easily find the proper address where to obtain either the answer, or the definitive denial of answer, than the basic scientist in search of a profitable application of his ideas.

The question, however, is not whether it is possible to redraw here and there the boundaries between some given scientific and technological communities, or where to turn for consultation. There have been, and there probably will be in the future, cases of instant match between science and technology. But the restriction of research to such rare instance of immediate technological relevance would slow down the growth of scientific knowledge and make research extremely inefficient. Obvious leads to discoveries which might revolutionize broad fields of knowledge would be passed over for minor technological solutions, because there would be no professional scientists to pursue leads of merely theoretical significance. In the long run this would also be economically inefficient. Important theoretical advances usually have great

3. For an analysis of the isolation of the community of medical researchers from that of medical practitioners, see Aran and Ben-David (1971).

indirect practical potentialities which can be exploited if there are entrepreneurs bringing them to the attention of technologists and others who may be concerned. But technological solutions, even if they have great potentialities for generalization, will hardly reach that stage in the absence of professional scientists whose job is to theorize and generalize.

With the exception of countries which are content to 'parasitize' the results of fundamental research performed elsewhere[4] the optimal way to increase the uses of science is, therefore, not to select projects according to their supposed promise of applicability, but to increase the motivation and the opportunities to find uses for science, and to find practical problems which can stimulate research. The relationship between fundamental and applied research should not be visualized as a series of separate links between certain fundamental discoveries and their 'applications'. Rather, practical uses of science should be conceived as the result of chance interactions between fundamental discoveries on the one hand and practical interests on the other, which can occur in an infinite variety of ways. The purpose of policy should, therefore, be to influence the likelihood of these chance occurrences by increasing the density of both kinds of activities and the velocity of the circulation of ideas and problems from both areas of activity in spaces which ensure, interaction. Increasing the density is a matter of investment, velocity is the result of entrepreneurship, and creating the properly enclosed spaces is a task for organization.

These, by the way, are not only the means for a utilararian, but also for an idealistic, science policy interested first and foremost in the augmentation of scientific knowledge for its own sake. Such a science policy would still have to find the resources to finance research and thus it would have to take an interest in the practical uses of science. It makes, therefore, little difference whether economic considerations or scientific knowledge is considered as the more important social value: both scales of preference indicate the same practical policies.

4. Which can be to some extent afforded by countries from which migration of scientists and intellectuals is barred either by force or great national differences.

References

ARAN, L., and BEN-DAVID, J. (forthcoming), 'Socialization and career patterns as determinants of productivity of medical researchers', *J. Health Hum. Behav.*

CARDWELL, D. S. L. (1957), *The Organisation of Science in England*, Heinemann.

DENISON, E. F. (1964), 'Measuring the contribution of education (and the residual) to economic growth', in *Study Group in the Economics of Education, The Residual Factor and Economic Growth*, OECD, Paris, pp. 13–100.

HABER, L. F. (1958), *The Chemical Industry During the 19th Century*, Oxford.

HAGSTROM, W. O. (1965), *The Scientific Community*, Basic Books.

PRICE, D. J. de SOLLA (1963), *Little Science, Big Science*, Columbia University Press.

SCHMOOKLER, J. (1966), *Invention and Economic Growth*, Harvard University Press.

11 N. D. Ellis

The Occupation of Science

Abridged from N. D. Ellis, 'The occupation of science', in *Technology and Society*, vol. 5, no. 1, 1969, pp. 33–41.

The large-scale employment of scientists and technologists in government and in industry is a recent phenomenon. Most of the laboratories which are now an accepted feature of modern government and industry, have been established only since the beginning of this century. As Krohn (1961) remarks, the centre of gravity of science has shifted radically. The support of research by government and industry has grown to enormous proportions, and correspondingly, the traditional bases of support – university funds and private philanthropy – have declined in significance. Several writers (Bernal, 1939 and 1954, Wright Mills, 1956; Whyte, 1957; Wolffe, 1959) have argued that this change of location has been accompanied by certain attitude changes. Increasingly, the large scale formal organization has come to be accepted as the appropriate environment for research, and the lone independent investigator has lost his previous importance. Emphasis has shifted from the independent intellectual, pursuing knowledge in a context of maximum freedom, to the team directed along profitable lines. Also, research is increasingly justified (even by the universities) by its potential value in the achievement of human welfare, rather than by the value of a knowledge for its own sake.

Some statistics from a recent Government survey (HMSO, 1967) illustrate the extent to which science and technology are carried on in the industrial context (including both the nationalized and the private sectors). Data from this show that during the period 1964–5, over 60 per cent of the total research and development effort of this country (measured in terms of cost) was carried out in the industrial sector and accordingly, approximately 60 per cent of total active R and D manpower were deployed in this sector. During this period, the overall contribution of the universities to the total effort was only 6·5 per cent. Current expenditure by type of research activity was distributed thus:

12·6 per cent of the total expenditure was on basic research.
26·0 per cent of the total expenditure was on applied research.
61·4 per cent of the total expenditure was on development work.

Within the category 'basic research', 44 per cent of the effort was concentrated in the universities.

On the basis of this evidence, it is reasonable to suppose that the 'typical' work situation of the contemporary scientist (or technologist) is the industrial laboratory, and that his work is probably applied research and/or development work.

The knowledge – provided by historians and sociologists – we already have about the nature of the 'occupation of science' seldoms reflects this contemporary reality. Historians have tended to concentrate their efforts upon problems and events occurring in past centuries. We know a considerable amount about the early proceedings of the Royal Society and other esteemed scientific institutions of the past; yet little is known about the origins and development of the industrial laboratory – the most typically modern institution of science. Likewise, sociologists have debated at length about the nature of the social organization of 'pure' science, devoting much of their effort to defining its distinctive ethos. (See Merton, 1949, chs. 15 and 16 and 1963; Hagstrom, 1965; Storer, 1966.) It is therefore a relatively simple matter to read about the activities and thoughts of outstanding scientists of the past and the sentiments of pure scientists today. However, if we wish to consider the large majority of scientists and the largest sector of modern science – industrial scientists and technologists and their work – the literature is, as yet, limited. Very little is known about the present situation of the industrial scientist in Britain, and about how this has changed during this century.

Historians and sociologists have tended to avoid problems of this kind by excluding applied science, and those engaged upon it, from their discussions. Their definitions of 'science' and the 'scientist' are often based upon an assumption that 'pure' research constitutes 'real' science, and that the academic environment is the most natural one for the scientists. Other kinds of knowledge-seeking activities based upon the established scientific and technological disciplines are seen as being at their best, poor substitutes for the real thing. It has been widely assumed that many industrial scientists are 'itching' for an opportunity to do 'real' science – pure research – and that creativity is seriously restricted in an environment which does not provide opportunities for work of this kind. Thus the industrial scientist has often been portrayed as the 'aspiring academic', who has through some misfortune, wandered into the wrong camp! Evidence from recent studies and my own survey suggests that this view is very far from the real situation. The large majority of scientists whom I interviewed expressed a definite preference for applied research and did not regret being unable to pursue fundamental knowledge in their fields. (These remarks must be qualified by an acknowledgement of some recent studies;[1] the authors of these have questioned this traditional view of the scientist, and have provided some evidence to support their arguments.)

My own discussion involves rather different definitions of 'science' and the 'scientist' from those mentioned above. Applied research and development work are included under the rubric 'science', as is 'pure' research. Accordingly, those engaged upon applied science have as equal a right to the title 'scientist' as has the academic. If these other kinds of research are included, then the traditional distinction between scientist and technologist loses much of its supposed significance. In many of the laboratories I visited a distinction between the scientist and the technologist, according to the subject of the degree, had very little meaning; graduates of the 'pure' sciences and of technologies were often deployed on the same problems in multi-disciplinary research teams. Thus the term 'industrial science' may be taken as referring to the whole spectrum of research and variety of disciplines found in this sector.

The orthodox model of the industrial scientist

Many previous studies of the industrial scientist have taken as their central theme the conflict between two cultures – that of

1. Several writers may be mentioned: Robert Avery; Stephen Box and Stephen Cotgrove; John R. Hinrichs; Norman Kaplan; Roger Krohn; S. S. West.

'management' and that of 'science'. These writers[2] have assumed that this conflict characterizes many of the problems which occur in the administration of the industrial research laboratory. These two cultures – or 'ideologies' as they are sometimes termed – imply opposing sentiments about what is valuable and worthwhile. Several texts from the Harvard Business School, intended to provide a conceptual framework for administrators of industrial research, are explicitly based on a model of this kind. In one such study, Hower and Orth (1963) introduce a series of case-studies with an outline of this cultural conflict:

Here we are concerned with the divisions between the scientific culture and the management culture . . . the man who favours the values and ideals of the scientific community has an orientation predominantly external to the company he works for. Getting ahead, for him, means impressing fellow scientists – especially the recognized authorities outside his company – by his technical achievements. The man who identifies with the management community will usually, but not invariably, have an internal orientation.[1] (p. 31.)

Hower and Orth proceed to spell out these two very different cultural configurations. That of management includes several common assumptions; viz. – that a company is in business solely to make money; that some sort of management control must be exercised over each component of an organization if the company is to be effective in reaching its goals; that the work of each individual must be supervised by someone.

In general the management culture may be said to place a high value upon financial soundness; hierarchy authority; loyalty to the company; conformity with established policies and procedures; growth in business volume and in size of organization; 'getting action'; 'getting ahead'; and tangible private rewards (promotion and increased pay) for superior performance. (p. 35.)

The 'scientific culture' involves beliefs and values in direct opposition to these; including such sentiments as – freedom to

2. The following are broadly representative of the many works which could be cited: Hower and Orth III (1963); Orth III, Bailey and Wolek (1965); Kornhauser (1962); Marcson (1960); Princeton Opinion Research Corporation (1959); American Management Association (1958); Raudsepp (1963).

work on projects of the scientist's own choosing in his own way; freedom to communicate with scientists in other institutions; freedom from having to account systematically for time and money spent or results obtained; the need for independent verification of conclusions.... Hower and Orth then proceed to examine a series of case-studies which confirm the truth of these initial observations.

The work of Kornhauser (1962) presents us with a very similar view of the industrial scientist. Whereas Hower and Orth simply offer a number of rather bald statements about the motivation and attitudes of the industrial scientist, Kornhauser attempts a sociological explanation of these phenomena. His model is based on the premise that the industrial scientist identifies with a 'professional' colleague community external to the organization in which he works; this involves an 'ideological' commitment on the part of the scientist which conflicts with the 'bureaucratic ideology' of the management. Although Kornhauser's terminology differs from that used by Hower and Orth, his description of these 'ideologies' (professional and bureaucratic) is quite consistent with the two-culture model.

Kornhauser develops his theme by examining the mechanics of this conflict; he describes four spheres in which it can occur.

Goals

'Strains between science and organizations emerge first of all in the formulation of research goals.' (Kornhauser, 1962, p. 16.)

The 'professional community' and management have very different views about what is important and worthwhile scientific research. The scientist – being a member of this community – favours basic research goals, free from the constraints imposed by commercial objectives.

Scientific institutions seek to instill in their members a commitment to the growth of science as a discipline of the human mind, and an obligation to sustain the integrity of science itself. Individual scientists are not equally socialized to the values and norms of science any more than they are equally well trained in the technical skills of science. Nevertheless, scientists generally bring to their work an internalized set of standards for scientific activity. By virtue of their training they expect to contribute to science in their sphere of special competence. (p. 18.)

This statement contains the supposition that industrial scientists have generally internalized the values of 'pure' science during their undergraduate training. In addition, it is further assumed that these values and beliefs are sustained in the industrial context by the 'professional community'.

Scientists and managers come to have very different interests in research; management's goals for the laboratory differ from those its participants bring to it. A quotation from Kornhauser's text succinctly describes this dilemma:

> ... the issue of basic versus applied research expresses the underlying tension between professional science and industrial organization. Professional science favors contributions to knowledge rather than to profits; high-quality research rather than low-cost research; long-range programs rather than short-term results; and so on. Industrial organization favors research services to operations and commercial development of research. These differences breed conflict of values and goals; they also engender conflicting responsibilities and struggles for power. (p. 25.)

Controls

Conflict in this sphere stems from the interaction – in the context of the industrial laboratory – of two conflicting principles of control, one based on the principle of hierarchy and the other upon that of colleagueship. In the bureaucratic organization – i.e. the research laboratory – authority is vested in a series of offices that are hierarchically ordered. In contrast with this, 'professional' control is exercised by a company of equals. Thus, whereas scientific institutions are based on various kinds of colleague authority, government and industry rely primarily on administrative responsibility.

Incentives

Since the 'worlds' of learning and commerce hold very different views about what is valuable and important, each will offer its own kind of incentives for the industrial scientist.

> The scientific profession seeks contributions to knowledge by soliciting research papers for professional meetings and journals, and by rewarding intellectual excellence with honors and esteem. The industrial firm seeks contributions to production and sales by soliciting new or im-

proved devices, and by rewarding commercial success with promotions, in a hierarchy of status, income and authority. (p. 156.)

Influence

Kornhauser examines several sources of resistance to the utilization of research findings and the influence of research personnel. The difference between the *weltanschauung* of the scientist and that of management and other groups in the parent organization, may contribute to the isolation of the laboratory. Barriers to communication between research staff and other groups in the organization may result from a 'vicious circle' in which all parties adopt the view that a minimal level of contact is the happiest resolution for all concerned. Consequently the *weltanschauung* of each party will become firmly established and both sides come to define each other as 'hostile' and 'alien'.

Some limitations of this model

This model certainly expresses the contrasts between those sentiments which characterize 'pure' or 'adademic' science and those of industrial management. The objectives of industrial science are ultimately 'commercial'; the support given to research is motivated by a belief that it can contribute towards the achievement of certain economic goals. Thus, any scientific or technological advantage made by the research staff is finally subjected by management to an evaluation according to economic criteria. There is an enormous difference between this 'pragmatic' problem-solving activity and that usually associated with academic research. Both the initial conceptualization of the problem, and the criteria of adequacy applied in the evaluation of a solution are very different.

This model does, however, go much further than this. It does not simply assert the truism that there is a vast 'ideological' or 'cultural' gap between the 'worlds' of 'pure science' and 'commerce', but also that the majority of industrial scientists are integrated into the former, and are alienated from the latter. Several suppositions are used to justify this maxim; these have been investigated in some recent studies and my own survey.

The assumption that a significant proportion of science students come to internalize the 'ethos of pure science' during their undergraduate training has already been radically modified by findings from a study by Box and Cotgrove (1966). Their evidence shows that a significant proportion of science students have an instrumental view of science; they acquire the knowledge and skills of science and do not seek public recognition as scientists, being prepared to use their expertise for occupational advancement and to abandon it when expedient. Evidence from another study, that of Avery (1960), shows that even if the young graduate carries the 'pure science ethos' with him to the industrial laboratory, this is soon discarded. A process of enculturation occurs during which the scientist learns a new culture and comes to adopt it as a necessary ingredient of his working life.

Table I (Appendix I) provides comparable indices of 'importance to overall work satisfaction' for a series of items. High scores on items – C, E, H, I, and L – are associated with a 'pure science ethos'. The high scores for these items obtained from the sample of academic scientists certainly corroborates this. Academics, judging from both their scores on this battery of items and their answers to questions in the interview, undoubtedly attach considerable importance to being free to choose their own research problems and thus do fundamental research. In contrast to this, industrial scientists seldom emphasized the importance of this kind of freedom. Their scores on these items (C, E, H, I, L), together with their answers to interview questions, clearly showed that very few industrial scientists embraced the 'pure science ethos'; irrespective of whether they may have held these sentiments when they left the university. Most of those interviewed expressed a definite preference for applied research; the reason most frequently given for this was that applied science provided more intrinsically satisfying results – a tangible end product – than those usually associated with basic or fundamental research.

There is no doubt that most industrial scientists accept the framework of industrial science. Only a few of those interviewed expressed any principled objection to the commercial restriction of scientific publication. Most of the industrial scientists expressed the view that a policy of restricting publication was justifiable. Furthermore, most industrial scientists were not concerned about developing a reputation external to the organization through publications.

It was apparent that many industrial scientists were unaware that 'pure' science might have a distinctive moral ethos. Each interviewee was asked whether the academic who was faced with a choice between publishing or selling the results of his work, had a moral obligation to make his knowledge public to his fellow scientists. Most industrial scientists felt that he should follow his immediate self-interest; e.g. do what was best for his future career. Those who did imply that there was a moral issue involved, usually argued that if he was supported by public money the results of his research should be public knowledge. It was very significant that many of those answering this question just could not comprehend why it was being asked; this clearly showed that most industrial scientists – far from having internalized it – could not even discern an ethos of 'pure' science!

This model involves a further supposition of doubtful value. It is assumed that there is a vigorous level of collective organization among industrial scientists; that professional bodies function in science as effective sources of social control upon the industrial scientist.

As far as collectivism is concerned – formal associations and collective identities – there is a vacuum in British science. Many of those interviewed were unable to offer a title for their occupation – other than the general nomenclature attached to their position in the laboratory – and, as we would expect, did not identify with an occupational or professional grouping. The level of collective organization in British science (and to a lesser extent in American science) in any form – 'professional' or 'trade union' – is generally very low; one of the most striking features of science when it is compared with analogous occupations – such as journalism, law or medicine – is precisely this absence of a 'professional consciousness' and effective collective organization.

This brief assessment of the orthodox model may be concluded with a general comment. The model is based upon a view of science as a socially-integrated occupation. A reverse view – that of science as an amalgam of many diverse elements – will be developed in the remainder of this paper. The most distinctive feature of this occupation is the diversities it encompasses. In fact, when we attempt to study it, its amorphous nature is so apparent that it seems almost misleading to refer to such entities as 'industrial science' or 'industrial scientists'. The ideas presented in the next section may appear less ambitious than those of the model discussed above. This is inevitable, since my central conclusion is that holistic generalizations of any kind about the nature of the industrial scientist have a limited heuristic value.

The occupation of science

It is always a more difficult task to discuss something which is fragmented and amorphous in character, than a subject which has, or is assumed to have, some kind of unifying structure. To simplify this discussion I have developed two dimensions which provide a basis for a systematic survey of the structural features of this occupation. These two dimensions refer to the nature of the work roles performed by the scientist: a distinction is drawn between the *content* of these roles, and the *context* in which they are situated.

An occupation in which the work roles are almost homogeneous with respect to content *and* context has the structural prerequisites for collective organization, and for a common identity among its members. Furthermore, an occupation – such as medicine – which has these structural features, has a strong vocational emphasis in the training of new recruits. In those other occupations where these structural characteristics are not present, little basis exists for this kind of superstructure. Collectivism, in any form, implies some sphere of shared experience and/or interests on the part of those involved.

The nature of the occupational roles filled by science and technology graduates affects the educational structure. The two structural dimensions define broad limits within which departments of particular disciplines are able to develop courses based upon vocational rather than academic lines. If these dimensions are seen as dichotomies, then it is possible to have a four-fold classification of disciplines according to the nature of the occupations their students usually enter. (In fact there are only three groupings of subjects/disciplines, since there is a fourth type – where the content of the work roles is heterogeneous yet the context is homogeneous – which is just a logical category with no empirical instances.)

Vocational departments

The medical school provides an obvious example of a vocational department. The work roles of the practitioner are relatively homogeneous in both content and context; it is therefore possible to provide an education which is specifically directed to meet the 'needs' of these roles. Although the medical profession is differentiated and fragmented in some respects, there is nevertheless a large central core of general practitioners whose roles are homogeneous in structure. This central core of the occupation functions as a social object, providing a focus of identification for the majority of medical students. Furthermore, other segments of this profession – the physicians and surgeons – are also readily observable to the student. A process of anticipatory socialization occurs: thus, although the student cannot fully appreciate and know what it means to practise medicine until he actually does so, he is able to anticipate and know with some accuracy the nature of his future work roles (content *and* context) before he accepts the responsibilities of practice.

This contrasts vividly with the situation of the pure science student in relation to his future employment; even if he does try to define his education as a vocational training, the range of occupations which he can come to know well and thus identify with, is very limited in comparison to the variety of employment opportunities that exists. From his own personal experience he has come to know well the 'worlds' of education and learning and may therefore be tempted to identify with these occupations, rather than those other occupations about which he can know very little.

Departments based upon Industrial Technologies

These departments provide a training in a specific expertise associated with a particular industry, or with a range of processes common to several industries. There are many examples; agricultural technology, chemical engineering, colour chemistry, food technology, mechanical engineering, metallurgy, and textile technology.

The curricula of these departments are often closely related to the expertise required in particular occupational roles. However, the socialization of the student cannot be as complete as that we associate with medical education. Because the contexts in which technologists work are so varied it is not possible for the student to readily identify with his future occupation. Nevertheless, judging from my own study, technology students are more adequately prepared for the industrial environment than pure science students, most of the staff of these departments have spent some time in industrial employment before taking up their academic posts. Furthermore, there are often close links between these departments and their respective industries; many departments are situated in close geographical proximity to major centres of their industry and have established – through consultancy, sandwich courses, and past students – strong ties with local firms.

Departments based upon Academic Disciplines

These departments are based upon traditional academic divisions of knowledge, i.e. botany, chemistry, mathematics, physics, zoology. The graduates of these subjects are deployed in a wide variety of occupational roles – heterogeneous in both content and context. Thus their work is often unrelated, or only indirectly related, to the curricula of their undergraduate studies, e.g. the physics graduate who enters industrial research probably finds that the craft skills and technology of the industry are far more relevant to his work than physics.

A formal qualification in one of these disciplines may have a similar function to the 'arts' degree. The future employer often defines the pure science degree as an evaluation of present (and potential) intellectual development; not as a certification of the specific expertise required for a particular job. Thus the pure science graduate who enters industry finds that he is learning and applying an expertise which is quite different from that he learnt during his academic education. Since pure science graduates are dispersed in this way it is not possible to devise vocationally-oriented courses which could cater for so many diverse occupational roles; in any case, even if such courses were developed in these disciplines, the fundamental side of subjects would often be dropped and the essence of them lost, resulting in the curricula being hardly distinguishable from their related technologies.

There are some important exceptions to these general comments. Some pure science graduates are able to pursue occupations where there is a direct continuity with their undergraduate or postgraduate studies – fundamental/basic research in government laboratories (and some industrial laboratories) or the fields of higher education (universities, polytechnics, colleges of technology).

The scientist's own definition of his situation

The orthodox model presents a 'stereotype' of the industrial scientist; the person who is dedicated to the pursuit of fundamental knowledge in his discipline. However there is a danger that another 'stereotype' with opposite attributes which would prove equally inadequate, may be accepted in its place. Those who postulate a dichotomy between the 'scientific' and 'managerial' ideologies have tended to pose the wrong kinds of questions about the social relations of the industrial research laboratory. Thus, although it is certainly true that most industrial scientists accept the logic of applied science, this does not necessarily mean that they accept management's definition of their role. My own survey showed that although there are many conflicts between scientists and management, these rarely had any connection with the 'science' and 'management' conflict of the orthodox model.

If the industrial scientist is neither the 'aspiring academic' nor the 'professional' what is he? Certainly he is an employee in the large scale organization, and has all the usual problems associated with the position of subordinate to management. Most of the scientists I interviewed saw management (particularly higher management in the parent organization) as something quite apart from themselves. The subordinate position of the laboratory worker often excluded him from participation in decision-making. Higher management in the parent organization often operated with an image of the research scientist as someone who could only be trusted with the technical aspects of a problem; thus his duties could be defined very narrowly, excluding him from a wider involvement in the company's operations. He was often seen as being incapable of comprehending the subtleties of commercial policy and its relevance to his own work.

On the other hand, many scientists looked for a greater involvement of this kind; they wanted opportunities to acquire a wider experience of the company's operations and greater responsibilities; this they felt, would provide a broader context for their technical work. Their reasons for this were varied. For some, a wider experience of this kind was seen as a possible basis for moving away from the laboratory into other departments of the company. However, for many others it was not linked with a desire to leave research, it was simply a feeling that this wider perspective could give their technical work a greater sense of relevance. This lack of 'meaning' or 'relevance' for their technical work, was frequently mentioned by scientists as a source of considerable dissatisfaction with their present work situation.

My interviews with scientific civil servants showed that a similar situation exists in government laboratories. The Institution of Professional Civil Servants (IPCS), in its evidence to the Fulton Committee, draws attention to this problem and recommends that the scientist's position in the Civil Service hierarchy should be changed:

There is an urgent need to implement ... the basic principles of good organization: responsibility should be placed on the man doing the job. If the man is in charge of a project or of a continuing job, he should be responsible for seeing that it is done efficiently. Responsibility implies power – power to allocate staff to organize and to manage them, to plan work and, above all, power over the purse strings. ... the professional officer, like others, should be expected to act within his judgement – and to take the consequences if his judgement is wrong. (IPCS, 1967, pp. 7–8.)

The scientific civil servant's present position is described in the following terms:

The implication is that once an activity, however specialist, involves more than a relatively small expenditure or slight contact with the public it automatically becomes administrative in character. It is not apparently envisaged that any professional (civil servant) can ever be capable of weighing political, financial and economic considerations. (IPCS, 1967, p. 57.)

This definition of the scientist's role – the expert on the side-lines without any real managerial responsibilities or authority – has a wide traditional acceptance in government and industry.

Finally, it is necessary to mention some important differences I observed between the situation of the 'pure' science graduate in industry, and that of the technologist. I have remarked previously that the technologist has usually undergone a training which prepares him fairly adequately for his work in industry. On the other hand, the science graduate often finds himself in a very different position; his academic training does not usually provide a specific expertise which is immediately relevant to his work in the laboratory. The scientist finds himself having to learn a relevant expertise and competing from a disadvantageous position with the technologists around him. A young science graduate – judging from my interviews – can find it difficult to adjust to this situation and it is likely that he will feel under-utilized in his new environment. The low score for 'present level of satisfaction' on item G, appendix I, 47 for scientists, compared with 65 for technologists, clearly shows that this feeling exists. It must be added that both groupings are very concerned about being fully 'utilized' in their work, their respective scores of 94 and 85 illustrate this.

It may be added that the technologist is generally far more certain about his role in the industrial context, experiencing little (if any) ambiguity about his self-image. As a result of his training he comes to define himself as an industrial practitioner; furthermore, his occupational title and 'professional institutions' have a legitimate place in industry – e.g. the mechanical engineer's long-established identity. The technologist is therefore able to concern himself with developing a reputation among his fellow technologists, and with pursuing a career within research and development. On the other hand, the scientist can no longer identify with his discipline (and probably does not even want to); at the same time, however, having started from a disadvantageous position vis-à-vis the established technologist, he finds it difficult to readily adopt his new identity – that of the technologist. His situation may appear to him as 'anomic' – no longer belonging to 'science', and intruding on the province of others, if he enters his newly acquired technology.

Some scores in Table I substantiate these comments. A very significant pattern emerges: the industrial technologists scored consistently higher than industrial scientists on those items – E, F, H, I, L, – which may be taken as indicators of a commitment to a career in their specialisms. At the same time, however, scientists are generally more concerned about gaining experience in administration, which might suggest that at least some of them are looking away from the laboratory for a solution to their present 'anomic' situation.

To conclude, it must be stressed that the 'marginality' experienced by some industrial scientists does not derive from the persistence of a 'pure science ethos'. These scientists are not 'marginal' because they seek to be part of a 'world of pure science'; nearly all those I interviewed expressed a positive inclination for applied research, and accepted the logic of industrial science. Far from wanting to withdraw, many sought a greater involvement in the 'world of industry'. The 'anomic' situation in which they may find themselves, is due to circumstances beyond their own volition. The funnelling process towards orthodox academic achievement, which usually starts early in the secondary school career, tends to present only one set of alternatives – the pure sciences. The technologies are seldom presented to the pupil as worthwhile alternatives; they are often seen as subjects suitable for the less able pupil. Often the industrial scientist who expressed a sense of 'anomie' about his situation, was well aware of being a victim of this process, regretting that he was not presented with the opportunity of pursuing technology at a much earlier age.

Appendix I Dimensions of work satisfaction

Comparable indices for scientists and technologists in industrial research and universities

1. Question answered by respondents: 'listed below are a number of conditions and auxiliary services that may be attached to your position in the organization. Could you please answer *two* questions about each condition? (please ring the appropriate number.)
   Firstly: how important do you feel that item is for your *overall work satisfaction*?
   Secondly: how satisfied are you at present with each condition?
2. In the table below, the first figure in each box is an index of 'importance for overall work satisfaction'. The second figure (in brackets) refers to 'present level of satisfaction'.
3. All indices are within a range 0–100. A high figure indicates that the item is of high importance to work satisfaction (or that there is a high level of satisfaction).
4. Indices are adjusted scores arrived at by method of summated ratings.
5. It must be stressed that these statistics only permit general comparisons between classes and between items within a given class. The industrial research sample does not include government laboratories, or research associations.

Item	*Scientists*		*Technologists*	
	University	*Industry*	*University*	*Industry*
A Salary	55 (31)	77 (50)	81 (40)	76 (52)
B Quantity and quality of assisting personnel	79 (47)	78 (38)	65 (40)	63 (39)
C The amount of free time available for private research	90 (74)	38 (49)	73 (66)	38 (65)
D Opportunity for gaining experience in administration	20 (64)	63 (37)	32 (64)	55 (41)
E Prestige of this department in the scientific/technical world	60 (51)	43 (50)	67 (37)	51 (60)
F Prospects for promotion up a research career ladder	60 (49)	75 (44)	70 (38)	91 (54)
G Extent my qualifications and experience are fully utilized	81 (84)	94 (47)	95 (70)	85 (65)
H The opportunity to pursue basic research in my field	90 (83)	33 (60)	70 (85)	49 (65)
I Freedom to choose my own research projects	88 (96)	51 (49)	80 (85)	54 (56)
J The degree of freedom I have to manage my own work	96 (93)	91 (63)	88 (93)	90 (79)
K Opportunity to attend scientific/technical meetings/ conferences	72 (60)	65 (51)	70 (45)	60 (67)
L Opportunity to work with highly reputed technologists/ scientists	63 (77)	44 (49)	72 (38)	50 (61)
N =	50	118	40	75

Appendix II Some notes on the survey

This survey is intended to provide part of the material being collected for a Ph.D. dissertation in Sociology. In the course of the field work I have visited about thirty research laboratories and numerous university departments.

The final survey – conducted by means of a questionnaire completed prior to a forty-five minute interview – is being carried out among scientists and technologists in about sixteen research laboratories and about sixteen university departments. The industrial sample included among others, a government laboratory, a nationalized industry laboratory, a research association, and an independent industrial laboratory. The laboratories visited for the survey are situated in a variety of industries: including chemicals and dyestuffs, electrical engineering, electronics, food, mechanical engineering, textiles, tobacco manufacturing.

Altogether a total of about 400 scientists and technologists have so far been interviewed. At the present time, since the survey is still in progress, the distribution of the sample can be only given in general terms. In the university sample, about fifty scientists from departments of the biological sciences, chemistry, and physics, have been interviewed. A similar number of technologists from engineering departments are also included. In the industrial sample, about 175 scientists and 125 technologists have been interviewed.

References

American Management Association (1958), 'The management of scientific manpower', Management Report 22, New York.

AVERY, R. W. (1960), 'Enculturation in industrial research', *IRE Transactions on Engineering Management*, vol. 7, March.

BERNAL, J. D. (1939), *The Social Function of Science*, Macmillan Co.

BERNAL, J. D. (1954), *Science in History*, London.

BOX, S., and COTGROVE, S. (1966), 'Scientific identity, occupational selection and role strain', *Brit. J. Sociol.*, no. 28, March.

HAGSTROM, W. O (1965), *The Scientific Community*, Basic Books

HMSO, (1967), 'Statistics of science and technology', Dept. of Educ. Sci., and Min. of Tech.

HOWER, R. M., and ORTH III, C. D. (1963), *Managers and Scientists*, Harvard University Press.

IPCS (1967), 'Evidence to the committee on the civil service', vol. 67, pp. 7–8.

KORNHAUSER, W. (1962), *Scientists in Industry*, University of California Press.

KROHN, R. (1961), 'The institutional location of the scientist and his scientific values', *IRE Transactions on Engineering Management*, vol. 8, September.

MARCSON, S. (1960), The Scientist in American Industry, *Industrial Relations Center*, Princeton University.

MERTON, R. K. (1949), *Social Theory and Social Structure*, Free Press.

MERTON R. K. (1963), 'The ambivalence of scientists', *Bull. Johns Hopkins Hospital*, vol. 112, pp. 77–97.

ORTH III, C. D., BAILEY, C., and WOLEK, F. W. (1965), *Administering Research and Development*, Tavistock.

Princeton Opinion Research Dorporation (1959), 'Conflict between management mind and scientific mind'.

RAUDSEPP, E. (1963), *Managing Creative Scientists and Engineers*, Macmillan Co.

STORER, N. W. (1966), *The Social System of Science*, Holt, Rinehart & Winston.

WHYTE, W. H. (1957), *The Organization Man*, Doubleday; Penguin (1960).

WOLFFE, D. (1959), (ed.), 'Symposium on basic research', AAAS, Washington.

WRIGHT MILLS, C. (1956), *White Collar*, Oxford University Press.

Part Four

Science and Political Institutions[1]

The increasing resource-needs of science and, more important, increasing political consciousness of the value of research have, over the century, produced extensive links between science and political institutions. Intermediary roles and institutional forms have proliferated, mostly around the activity of channelling resources into science. But, as Yaron Ezrahi points out in the first paper of this section, scientists do not simply compete for resources with other general sectors of expenditure; they also compete with each other in so far as they belong to different disciplines and specialties. External evaluations, and hence external images, of particular disciplines now decisively affect their internal rates of growth. Ezrahi describes how elements of these images can be manipulated as 'political resources', and does not refrain from indicating how such manipulation can itself have internal consequences.

Interaction between science and politics has increased most markedly since the second World War. Scientific participation in the war effort particularly through the Manhattan Project, the development of radar, and Operations Research had been notably successful; scientists enjoyed prestige not simply as experts, but as individuals specially equipped by their scientific training with general skills applicable to a wide range of problems. In the United States, it was expected that they would increasingly fill political roles as decision takers and policy makers, as well as experts and advisors. Optimistic analysts expected them to revolutionize the operation of political

1 For an analysis of the relationship between science and politics at an altogether deeper level of abstraction see the final reading of this volume, and also Habermas (1971).

institutions. But optimism declined as scientists accepted the political responsibilities expected of them. It became apparent that the value conflicts at the heart of politics divided scientists as much as the wider society, and indeed affected their judgement in concrete situations; away from their areas of particular expertise, scientists could it seemed be as fallible as anyone else. (See Wohlstetter, 1964.) So the tendencies of the mid 1940s have not led to significant institutional changes in the political area, although the links between science and politics have continued to grow. In Britain and Europe of course, scientists never came to walk the corridors of power in such large numbers.

Relations between science and politics have involved conflict as well as cooperation, usually as a result of political initiatives. Historically, such episodes have become memorable when they have involved particularly distinguished individuals, Galileo in the seventeenth century, Vavilov and Oppenheimer in this. For the sociologist, they are important in that they generally represent attempts to impose external social control upon science, and make it possible to investigate the operational consequences of such control.

During this century the Soviet Union has provided a paradigm of science under overt political control. Functionalist sociologists in the United States long regarded Soviet research performance as confirmation of the value of scientific autonomy; the notorious dominance of Lysenko in genetics seemed to settle the issue,[2] and theorists ran, perhaps a little too easily, from the manifest harm done by a particular type of control in a particular instance, to general analyses of the dysfunctional nature of any kind of external direction. Since the launching of the Sputniks in 1957, sociologists have been noticeably less sweeping with their generalizations in this area, and more sensitive to its complexity. This was a desirable change, whatever opinion is taken of the reasons for it. It indicated the need to reappraise the status of theoretical analysis in this area in the light of more detailed and wide ranging evidence. Such a reappraisal we still await, although it is important to emphasize

2. For accounts of the effects of political control on Soviet science, particularly genetics, see Joravsky (1961 and 1970).

that no *a priori* reasons exist for assuming that it will involve major theoretical changes.

Occasionally, political conflicts may be generated by scientists themselves. At various times they have attempted to form professionally based pressure groups through which to exert political influence. Gary Werskey's analysis of such activity on the part of British scientists between the wars suggests that only unifying professional grievances can effectively maintain the cohesion of such groups. In periods when the prestige of science is high and orthodox political channels are open to its practitioners, the influence of such movements is unlikely to be significant. In developing his account Werskey provides useful insight into the low prestige of science among intellectuals and politicians of the 1930s, thus confirming the importance of the second World War in enhancing its status and influence.

References

HABERMAS, J. (1971), *Toward a Rational Society*, trans. J. J. Shapiro, Heinemann.

JOKAVSKY, D. (1961), *Soviet Marxism and Natural Science*, Routledge & Kegan Paul.

JOKAVSKY, D. (1970), *The Lysenko Affair*, Harvard University Press.

WOHLSTETTER, A. (1964), 'Strategy and the natural scientists', in R. Gilpin and C. Wright (eds.), *Scientists and National Policy Making*, Columbia University Press.

12 Yaron Ezrahi

The Political Resources of Science

Y. Ezrahi, 'The political resources of American science,' *Science Studies*, vol. 1, no. 2, 1971.

In the light of the traditional ethos of science, and particularly its emphasis on a complete separation between science and politics, a discussion of the political resources of science is bound to appear somewhat perverse if not entirely heretical. Yet, while the idea of the separation of science and politics may still be consistent with feelings and wishes prevalent among scientists, it seems increasingly inadequate as a statement about reality. The tendency to confuse the question of what the social status of science *is* with what it *ought to be* continues to interfere with the development of a fruitful theoretical discourse about the social and political aspects of science.

Attempts to call attention to the links of science with its political environment have usually stimulated in scientists such associations as the Marxist approach to the sociology and history of science, which views scientific theories and institutions as the outgrowth of specific social and economic conditions. While some scientists have been attracted by this analysis, most have strongly rejected this type of approach, on the grounds that the description of science as derivative from its social context destroys its claim to be politically neutral and its right to be free from external control. It has suggested to them the dangers of censorship and political interference, and aggravated their fear of the consequences of linking science and politics. By the same token, scientists have tended to be much more sympathetic to those theories in the sociology and history of science which perceived scientists as obeying the inherent imperatives of scientific ideas and the logic of inquiry. The theory that scientists follow only the internal rules of science would seem to reinforce their effort to prevent the subordination of their work to standards extrinsic to science and to protect themselves from external political interference. Such autonomist social theories of science, as formulated by Polanyi, Hagstrom and Storer, provide, however, a more adequate description of the social reality as it was when science was still a relatively small enterprise, largely insulated from the mainstream of economic needs and political forces. Even then, much like classical economic theories of the free market system, the autonomist model of science could never substantiate the existence of a perfectly self-regulatory and independent market system of scientific ideas. But the fact is that scientific, like economic, activity was sufficiently differentiated from other social and political activities to have rendered such a theoretical perspective quite fruitful.

Yet, in the light of the increasing interpenetration between science and politics in the years after the second World War, an autonomist social theory of science has consistently failed to account for some of the most dramatic developments in the interrelations between science and society. By overlooking or dismissing the links between science and politics, such a theory was unable to explain the logic behind the growth and development of such bodies as the National Science Foundation (NSF), the President's Science Advisory Committee (PSAC), and the National Academy of Science-Committee on Science and Public Policy (NAS-COSPUP); or the increasingly influential role of the growing group of scientist-statesmen such as Vannevar Bush, Killian, Kistiakowsky, Wiesner and Brooks. To overlook such developments on the grounds that the mixing of science and politics is undesirable is, of course, to accept the unscientific practice of rejecting a statement about reality, not because it is proven false, but because it does not correspond with one's wishes. It is equivalent for example, to the suggestion that economic theory should not have readapted its conceptual apparatus to account for economic behaviour in which the government plays a growing role, because such political interference conflicts with the true values of *laissez-faire* economy.

Without the fallacy of mistaking the insular ethos of science for a theory about the actual place of science in society, it becomes easier to recognize that contemporary American science is not a socially autonomous enterprise, nor is it insulated from politics.

On the contrary, the unprecedented degree to which science in America is dependent upon external material and political support in order to exist has compelled American scientists to engage actively and continually in competition with other social groups for their share of public resources and political support. The new political condition of science has meant that the ability of science to grow and flourish depends no longer merely on the free and successful use of intellectual resources, but also on its adaptability to political action and its capacity to convert its unique resources into effective means of political influence.

However, the ability of scientists to form a disciplined group for effective political action in the public arena has been seriously constrained by the internal division of the scientific community into a multiplicity of specialized scientific groups.[1] If modern science had had a grand theory which unified all its parts into a single conceptual scheme, it would have probably been easier for the scientific community to organize itself politically and to discipline its parts in the name of an agreed concept of scientific priorities. But even though in the past the unity of all the sciences was widely considered a real theoretical possibility, it has remained primarily an ideal or a theoretical postulate. There have, to be sure, always been agreed criteria for the evaluation of the relative merits of scientific theories, such as their scope of explanation, predictive value, quantifiability and accuracy. But there has been no clear theoretical basis for a scientific order of priorities or status among scientific fields. In the absence of such internal standards that would be binding on scientists in different areas of science, each specialized scientific group would naturally be tempted to attach greater importance to its own theoretical objectives and methodologies than to the others. Nevertheless, as long as the internal process of science was socially autonomous and invisible to the public eye, the ability of any particular scientific area to grow and develop usually depended on the success with which it could demonstrate its scientific merits to scientists from different fields, and so acquire recognition within the scientific community generally. But when the once subtle and publicly

1. Price (1965, pp. 101–19), has pointed out that this handicap is one of the reasons why society has very little cause to fear the rise of scientific oligarchy.

invisible process of allocating intellectual and material resources among scientific fields became publicly visible and politically exposed, the intellectual justification of a scientific choice among scientists became inseparable from the political justification of that choice to the public. It is possible to argue that the success of any given field to mobilize material resources and social support has come to depend not merely on scientists' recognition of its intellectual merit, but also on its political, economic or moral appeal to laymen. Since the domain of politics is governed not by the attitudes and opinions of scientists but rather by the attitudes and opinions of the lay public, the resources of political influence available to any given scientific field depend less on what scientists think about it than upon how it is perceived by non-scientists. Though such lay perceptions of science may be regarded by scientists as inaccurate or utterly fallacious, they are political facts which have a great role in influencing the political resonance of science. In fact, recurrent myths and misconceptions associated by the lay public with science as a whole, or with any particular field of science, are important constituents of the social and political environment of science.

In view of the growing role of political factors in the development of modern science, the study of the particular ways in which science is perceived by the lay public and of the comparative public images of different scientific fields is pertinent not merely to the understanding of the interaction between science and society, but also to the understanding of the structure and direction of the internal development of many scientific fields.

In venturing into such a study, I would like to suggest that we distinguish within the different dimensions of science which are visible to the lay public and have political consequences for science, the following four categories of what may be called 'political visibility':

The relation and relevance of scientific pictures of reality or images of nature to prevailing social, political and religious beliefs.

The relation of technologies generated by different fields of science to prevailing social values and concerns.

The degree of accessibility of a given science to the public.

The degree of peer consensus among the scientists of any given field.

These categories of political visibility refer to the publicly perceived features of science which affect lay attitudes towards science; they do not, however, in themselves indicate whether these attitudes are positive or negative. Of course only the publicly visible traits of science which evoke in the lay public a sense of harmony between science and prevailing social beliefs can be regarded as political resources of science, whereas those visible traits which suggest a conflict between science and popular beliefs are its political liabilities. The political skill of scientists should therefore consist largely of the ability to exploit relevant social beliefs and attitudes in order to manage the public images of science so as to improve its positive political visibility and its capacity to evoke public support.

With respect to the first category, the pictures of reality or images of nature associated with particular sciences are not analysed here from the point of view of their explanatory or heuristic function inside science, but from the point of view of their external relations to prevailing beliefs. The social history of science provides some dramatic examples of the negative political consequences of the conflict between scientific pictures of the universe, such as the Copernican and the Darwinian, and deep religious and ethical beliefs. Pictures of the universe presented by physics and biology were similarly consequential for social attitudes towards these two sciences in different ages. Gillispie (1960, pp. 198–9) has pointed out that, to the Romantics, 'nature was the seat of virtue and Newton's laws were morally unedifying. . . .' They revolted against the quantitative abstractions of physics which they linked to the objectification of nature and the cosmic alienation of man. In their attempts to make a scientific picture of the universe more congruent with their concept of man, they turned to the qualitative sciences; and they later attempted to substitute biology for physics as the queen of science.

With respect to the social sciences, perhaps the best known example is the socialist objection to the concept of 'economic man' postulated by classical economic theory. Again, regardless of the strictly scientific utility of this concept in statistical and predictive operations, the notion that man is a calculating egotist, while consistent with the norms of liberal democracy, was largely unacceptable and therefore detrimental to the growth of economic sciences in some Communist countries.

A curious example of political taboo in the area of population statistics can be found in Lebanon, whose political system is based on the principle of a delicate balance between the Christian and the Moslem populations. Here a population census has been frozen for decades, since the lending of scientific certification to a picture of social reality incompatible with the fiction of balance between religious sects might have disruptive repercussions for the political system.

In as much as American pluralist democracy has not been a fertile soil for the growth of well articulated comprehensive ideologies, the impact of scientific pictures of the universe on public attitudes towards science has been less focused and more subtle, though by no means less consequential than in Europe. On the other hand, the unique American tendency – noticed by many students of American culture – to direct and judge human conduct in the light of empirical facts has rendered the capacity of science to authorize and certify facts and pictures of reality a potent source of political influence. Price (1962, p. 27) pointed out, in his pioneering study of science and government sixteen years ago, that in the American political system the 'unwillingness to take the answer from established authority leads to a tremendous use of research as a basis of decisions at all levels'. This clearly suggests that in America the reliance on scientifically 'certified facts' has been a matter of determining not merely the content of decisions but also their public credibility and legitimacy. In societies where social goals and policies are guided by concepts of transcendental authority, or are provided by revered aristocracies, the pronouncements of science about the universe, if allowed to be made public at all, would not usually have direct repercussions on human behaviour.[2] But in a society which has rejected such

2. It has been reported, for instance, that in Indian villages the sheer distribution of scientific information about modern agricultural techniques through television broadcasts failed to have a serious impact on the practices of the villagers. But when the new techniques were legitimated by the authority of the village chiefs in social forums following the broadcasts, they were more widely adopted. See Mathur and Neurath (1959).

models of hierarchical or elitist authority, almost any visible conflict between the premises of public policy and what are accepted as the objective and impersonal facts of reality smacks of arbitrary and abusive use of political authority. It is no wonder that the justification of decisions by reference to research or investigation committees has acquired in America a symbolic-ritualistic function similar to the medieval practice of linking important decisions to precedents and predictions from Holy Scripture.

The links between the authority of scientific certifications, and the public evaluations of government policy, have opened up great opportunities, as well as great dangers, for science in America. The newly found power of American scientists to determine the timing and the context of public scientific pronouncements about certain facts could have major political consequences – as the controversies over nuclear fallout, anti-ballistic missiles (ABM), food additives and the biological basis of racial differences can illustrate.

The recent controversy about the sources of differences in IQ distribution among different ethnic and racial groups is particularly instructive.[3] The environmentalist perspective on man and society has always been more compatible with the traditional American belief in egalitatianism than the hereditary approach. If differences in IQ performance among human groups are believed to be not hereditary but rather the function of environmental conditioning, visible inequalities can be accepted as a tolerable passing phase, and education can be regarded as a grand equalizer. In the context of the growing controversy about the racial aspect of educational and welfare programmes, a study challenging this environmentalistic model could only be regarded, by both supporters and critics, as enormously explosive. No wonder that the impact of the now famous publication of Jensen's article in the *Harvard Educational Review* was reportedly viewed by Washington policy makers as a major threat. The *New York Times Magazine* of 2 November 1969, quoted Special Presidential Assistant Daniel P. Moynihan as

3. For samples of the public record of this controversy and the rebuttals, see Jensen (1969); Congressional Records (1969a, 1969b, 1969c); *Life* (1970); *New York Times Magazine* (1969); *Bull. At. Sci.* (1970a, 1970b). See also my article (Ezrahi, forthcoming) and Bodmer and Cavalli-Sforza (1970).

saying that 'the winds of Jensen blow in this city at gale force', and admitting that the Jensen case was raised in a cabinet discussion. In the light of the great unpopularity of the hereditary approach to human intelligence and education, it is perhaps not a coincidence that the geneticists' community (and in fact, through the NAS, the scientific community as a whole) tried to dissociate itself from the linking of the genetic explanations for IQ distribution with educational policy.[4] The two most vigorous advocates of the implications of genetic factors in IQ distribution for educational policy were a physicist and a psychologist, while the majority of the geneticist community, which was motivated at least in part by anxiety about its public image and support, took great pains to criticize these efforts.[5]

The second category of public visibility of science concerns the relation of technologies generated by different fields of science to prevailing social values and concerns. The extent to which the links of science to specific technology constitute positive or negative political visibility depends on both the publicly perceived contributions of any given field to specific technology and the value attached to this technology by the public. The political visibility of science from the point of view of its links to technology would be relatively small in societies or cultures which reject the values of man's control over his natural environment or which do not recognize the links of technology to the conceptual dimension of science. It will also be reduced in societies where scarcity of financial resources limits the vision of technological possibilities. By comparison with those of other countries, the American system has been exceptionally prominent on both counts: first in its zest for the values of human control over the physical environment and the instrumental significance which it attaches to scientific conceptualization, and second in the availability of resources for the development of technologies. But the dynamism of American politics has rendered the political value of any specific links between

4. For an earlier version of this dispute within the American Association for the Advancement of Science (AAAS), see Price (1963 and 1965, pp. 110–11). See also Baker and Allen (1968, pp. 100–43).

5. These remarks refer to the *de facto* political repercussions of the 'heredity *vs.* environment' controversy; they are not intended to imply that the contending theoretical positions must necessarily entail the policy implications attributed to them by parties to the dispute.

science and technology vulnerable to frequent fluctuations in political orientations and public opinion. At the time and in the place where the climate of opinion was primarily that of conquest and development of nature, the links of science to industrial technology, through chemistry, entailed positive political visibility for the chemical community. But now that this climate of opinion has begun to give way to mounting concern over an 'ecological crisis', and growing suport for conservationist values, the public perception of chemistry as an ally of the values of industrial development, and of its role in the production of food additives, constitutes a political liability for the chemists' community.[6] On the other hand, those sciences which appear to have links to the goals of restoring and maintaining a balanced and humanly acceptable ecological system have only gained from this trend in public opinion.

Another case in point is the major shift in the relative positions of the physical and the social sciences in the last few years. As Harvey Brooks has pointed out, in the period between the last World War and the beginning of the Sixties the physical sciences enjoyed positive political visibility because of their links to military technology. During that period the social sciences underwent a difficult struggle for public recognition and support. The physicists' community, which was clearly the leader and the most influential group of scientists in public affairs, by and large did not support the social sciences in this effort and often resisted them vigorously. In November 1945, a letter to the President of the United States signed by five thousand scientists in support of the Bush Report stated 'that it would be a serious mistake to include the social sciences' (in the proposed NSF). It was widely held among physical scientists that, because the social sciences were 'controversial', their inclusion would render the NSF vulnerable to political criticism and would weaken its capacity to mobilize support for the physical sciences.[7] Since the mid-1960s, however, in light of growing criticism of the Vietnam war and the military, and signs

6. For a criticism of the role of chemistry in the food industry, see Ralph Nader's student project on food protection and the Food and Drug Administration. Turner (1970).

7. See the report by the Science Policy Research Division, Legislative Reference Service, Library of Congress (1969, ch. 5).

of profound social unrest, the contributions of the physical sciences to military technology have begun to boomerang, while their remoteness from urgent social problems has become a serious disadvantage. Although the social sciences have not developed spectacular means to solve social problems, the pre-occupation with such problems has been sufficient to give them a boost. Their status within the NSF, the NAS and the American Association for the Advancement of Science (AAAS) has visibly improved. Now that the competence of social scientists has become so relevant, and the social sciences are regarded as good company, the spokesmen of the physical sciences no longer seem ashamed to associate with the social sciences in public; and they even take the initiative in searching for common grounds with social science in coping with such environmental problems as pollution or the sonic boom. These examples clearly illustrate, I believe, the opportunities and the threats which are involved in the political visibility of the links between science and technology.

We have defined the third politically visible dimension of science as the degree to which it is accessible to the public. The role of the public accessibility of scientific knowledge in influencing lay attitudes towards science is not, of course, new. Francis Bacon criticized Aristotelian scholasticism on the grounds that it was inaccessible to the public. He was sensitive to the fact that, in an era of increasing challenge to established authority and the rising strength of anti-hierarchical values, his presentation of modern science as a new and more accessible mode of knowledge – not filtered by the esoteric mastery of books and Latin but open to the senses and common experience – served to bestow greater public legitimacy upon it. The emergence of the lay public as a legitimate audience of science was clearly manifested in the decision of men of knowledge, such as Galileo and Descartes, 'to write their works in the vernacular rather than in Latin avowedly for the purpose of appealing against the learned world to an intelligent reading public'. (See Butterfield 1966, pp. 180–81) In pre-revolutionary France, the exceptional popularity of qualitative chemistry among French democratic circles was similarly related, as Gillispie has pointed out, to the perception of chemistry as an exoteric science, as against the esoteric mathematical abstractions of Newtonian physics. (See Gillispie, 1960, pp.

184–6.) No wonder that in such an atmosphere the Secretary of the Académie Française, Condorcet, believed that the enhancement of the accessibility and visibility of scientific truths to the public was a necessary condition for public recognition of the authority of science in society.

In early nineteenth-century America, the popularity and the accelerated growth of what were then known as the 'natural history' sciences were linked to their ability to provide the common man with a sense of participation in the wonders of nature. The idea that the principle of scientific knowledge is classification was associated with the idea that science is not the work of geniuses and that 'everybody can be a scientist at least in comprehension'. (See Miller, 1965, p. 319.) As the experience of chemistry and natural history clearly shows, it is in its incipient stage that a science is most likely to appeal to the layman. When a science achieves a high level of conceptual development, it requires more elaborate skills and training, and thus becomes more professionalized and esoteric. In the context of the American populistic and egalitarian political values, the process of professionalization and specialization, though it may be a measure of success from the scientific point of view, involves considerable costs in terms of negative political visibility. There is in America a powerful political sentiment against any form of elitism or claim to exclusive authority or competence, whether political, religious or scholarly. Tocqueville (1945, vol. 2, p. 4), observing the American system in the middle of the last century, noted that 'in a country where no signs of incontestable greatness or superiority are perceived in any one of [the citizens] they are constantly brought back to their own reason as the most obvious and proximate source of truth'. Yet the most developed areas of science are usually the least accessible to the public and the most vulnerable to the charge of esoterism and remoteness. When a highly esoteric scientific field is also associated with a highly unpopular picture of reality or technology, the compounded negative political visibility may be particularly harmful. The fact that modern esoteric fields of science are often criticized on the same grounds on which the early propagandists of science criticized the exclusivity of the clergy suggests that, regardless of the content of knowledge, the modes of its social and organizational configurations influence the relation of the scholarly community to its socio-political environment.

Finally, the fourth politically visible feature of any specific science is the degree of consensus achieved by its member scientists. The political visibility of consensus was noticed long ago; Leibniz, for example, believed that controversies and conflicts of opinion among scientists reduced the social position of science, and he devised a demonstrative scientific encyclopedia in order to eliminate them. (See on this subject McRae 1961, Ch. 4.) Because the lay public cannot evaluate scientific propositions directly, it has to rely on more visible indirect signs of scientific merit, such as peer consensus. In the absence of such consensus, the lay public cannot find in science the certainty and the support it seeks. The ability of any specialized scientific group to make an impact on public policy and its implementation no doubt depends much upon this capacity to generate a minimal degree of consensus on scientific standards, evidence and conclusions, and to articulate them in the social context with the full backing of the authority of science. This is especially apparent in controversies such as on the relations between smoking or food additives and health, the nature of UFOs and the like, where the public is very anxious to receive scientific guidance, while the insufficient state of scientific knowledge limits the possibility of unequivocal peer consensus. When scientists can appear on all sides of an issue, none can persuade his audience that he speaks for objective and impersonal facts. The social force of scientific considerations is obviously weakened, and with it the public standing of the field of science that is involved.

The degree of consensus is obviously not uniform in each discipline with respect to all subjects. But there are differences in the theoretical basis of consensus among scientific disciplines. The physical sciences were notably more successful in this respect than the life sciences, and the latter more successful than the social sciences; and within the social sciences, economics showed the highest degree of peer consensus. A comparative, though admittedly impressionistic, look at the development and structure of the influence on public policy of the specialized scientific communities in these areas suggests most interesting correlations that have yet to be explored.[8]

8. I am currently studying the relations between such factors as the degree of theoretical consensus or conceptual differentiation of a field of science and

The four categories of political visibility of science which I have just presented (that is to say, the political dimensions of scientific pictures of reality, the links of science to technology, the accessibility of science to the public, and the degree of peer consensus) are hybrid variables. By combining the internal features of science with the traits of its socio-cultural environment, they can be used to conceptualize about the political resources of science. If we view science through these four categories, it becomes apparent that, in any given social context, fields of science differ in the character of their political visibility and resources; and that the political visibility of the same fields will vary in different political or cultural systems and at different points of time in the same system. It also appears that the same scientific field could hardly achieve high positive political visibility in the four categories simultaneously, since high scores in some will usually entail low scores in others. There is, for instance, very frequently conflict between positive political resources in the first two categories. Scientific pictures of reality which seem least consistent with common sense and popular beliefs are often most successful in predictive potential and in generating technologies. The mechanistic picture of the universe was historically widely unpopular on ethical and humanistic grounds, yet it was much more successful in predictive and technological productivity than the more ethically and humanistically popular biological vitalistic construct of the universe. Similarly, the scientific construct of the economic man was widely rejected on ethical and political grounds while supporting at the same time the development of fruitfully predictive and computational economic models. In the case of the apparent conflict between the hereditary and the environmental concepts of human intelligence, while the environmental model has enjoyed highly positive political visibility as an image of man, the hereditary model, because of the connotation of biological determinism, has suffered from negative political visibility. No wonder, then, that the supporters of the hereditary model have attacked the environmental model in the weak points of its failure when applied in educational

---

the extent to which it evolves a self-conscious group of professional workers with common perceptions of their relations to other fields of science and shared political resources and strategies for mobilizing the financial support and the legitimation of the public for their work.

programmes, while publicly defending the hereditary model not as an image of man but as potentially more applicable. (See the relevant sections in Jensen, 1969.)

These examples may reflect a more basic conflict between the force of scientific reductionism in predicting and technologically exploiting natural phenomena, and its unpopular fragmentary effects on common sense constructs of reality. The tension, in other words, is between the role of science as a cognitive enterprise which is a source of certainty and an integrated world picture, and as a source of power or tool for manipulating the environment.

There are also noticeable trade-offs in positive political visibility in the third and fourth categories. Very often the scientific disciplines with the highest degree of internal peer consensus on scientific matters are areas of science which are most esoteric and least accessible to the public. The greater the professionalization of a field, the greater is the exclusion of the layman.

The fact that scientific fields cannot usually achieve equally high scores in all four categories simultaneously has naturally led each scientific field to concentrate on emphasizing and utilizing its points of strength.[9] In the absence of a unified theoretical framework for science which would oblige the various fields from

9. This fact also helps to explain some of the patterns of relative rise and decline of the political fortunes of various scientific fields. Harvey Brooks, in comments on this paper, has pointed out that the decline of the political fortunes of physics and chemistry was not proportional to the rise in the fortunes of the social sciences and ecology, since, while physics and chemistry have been declining with respect to category two, they have maintained their usually high scores in category four; whereas social sciences and ecology, though they have gained with respect to two, have remained at their usual low with respect to category four. Indeed, the areas of the social sciences which have attracted growing support in Congress and the Executive and improved their status in the NSF, NAS, PSAC, etc., have been precisely those areas of the quantitative social sciences with relatively high scores in peer consensus (four).

Perhaps we should add the observation that, since, of the four categories of political visibility, gains in four often indicate progress also in terms of the internal theoretical development of science, and since scientists who are laymen with respect to areas of science outside their own expertise are nevertheless likely to be more sensitive than the public to the internal norms of scientific achievements, high scores in four are of particular significance for the endorsements of scientific fields by the larger scientific community, as represented by the NSF, NAS and PSAC.

within to conform to a scientific order of priorities, each scientific field has been tempted to use its unique political resources in attempting to mobilize public support for its particular endeavour. This freedom for independent political entrepreneurship has been particularly stimulated by the opportunities which have been opened to science during and since the second World War. The method of free political initiative has, however, involved serious potential dangers to science from the point of view of both its internal functioning and its general social status. Because federal funds now make-up an unprecedented proportion of the total material resources available to various fields, a system of uncontrolled competition for public support entails the possibility that the external test of political resourcefulness will dangerously outweigh the internal test of intellectual promise in determining the fate of different scientific fields.[10] The cumulative effects of the interpenetration of scientific and political criteria for the distribution of the scientific effort could cause serious imbalances and disorientation in the internal working of the scientific community. Such dangers are particularly acute because of the growing influence of the relative social images of scientific fields on the flow of young intellectual talent among them. The concern expressed by physicists, mathematicians and molecular biologists in 1969 and 1970 over indications of the declining student enrolment in these fields, compared with a rise in student enrolment in the 'relevant' sciences such as psychology, sociology and ecology, reflects these links.

From the point of view of the social status of science as a whole, such political *laissez-faire* entails the risk that scientists, by confronting the public with competing and conflicting claims, may erode the public credibility and authority of science as a whole. The public expects scientists to be unanimous, and is not inclined to accept the privileged authority of science when there is no socially visible consensus. The rationale behind Leibniz's concern is in this sense timeless. In the American context, the perennial need to justify claims for support before the public forums of Congress has, predictably, dramatized the political futility of a

10. Important circles of the American scientific community believe, for instance, that the space programme is a case of investment in a programme out of all proportion to its intellectual promise.

process in which representatives of competing scientific fields use all the arguments they have in store to present the relative importance of their own activities.

However, neither the internal nor the external deficiencies incurred by the practice of unrestrained political entrepreneurship were clearly detectable during the years (particularly after Sputnik) in which continually high public support allowed most scientific fields to grow at an unprecedented rate. But when this trend levelled off during the latter years of the Johnson Administration, and science was increasingly threatened by cuts of federal funds and deteriorating public support, the strategy of free and uncoordinated political competition began to show its weaknesses. It has now become increasingly clear that, politically and economically speaking, the resources mobilized by each field affect the reservoir of resources left for the others, and that the political tactics used by some fields affect the political options open to others. This is particularly true in areas of science remote from social concerns and government missions. Thus the fate of sciences with little political appeal may reach a crisis unless the political resources of relatively wealthy and publicly strong fields, as well as of the scientific community as a whole, are employed with greater economy and consideration for the overall state of science. Even though the NSF was specifically designated to be a sort of 'balance-wheel' which diverts federal funds to underfunded yet scientifically worthy fields, there was little that it could do to offset imbalances in cases where the meagre political appeal of scientific fields was greatly out of proportion to the size of the funds they required. Despite its specific efforts to enhance the weight of internal scientific considerations, the NSF has largely reflected the 'balance of power' among fields, rather than helping to modify it. The need to mobilize support for programmes from the lay public and its elected or appointed representatives, has forced the NSF to respond to the demands of effective political strategy and make concessions to extra-scientific preferences and criteria at the expense of more purely scientific considerations.[11]

11. The NSF has, to be sure, helped some scientific fields handicapped by low political appeal, such as systematic biology and pure mathematics, but it has nevertheless supported – somewhat out of proportion to its resources – fields with fairly strong political appeal, such as atmospheric science and oceanography.

In the light of the growing difficulties of the last few years, the feeling that a new political strategy is needed for science has gained considerable strength. These feelings have not led, of course, to a conscious planned shift, but they have produced some significant changes in attitudes and institutions, the full meaning of which cannot yet be evaluated. What seems to be emerging is a turn from the former system of basically free political competition to a moderately controlled competition in which each science is bound to use its political resources economically and in co-ordination with the interests of other sciences and the state of science as a whole. Some of the principles of this new strategy were echoed in the address of Professor Seitz, ex-President of the NAS, before the American Physical Society in November 1964. Seitz reminded his audience 'of the way in which the competition among the nations of Western Europe in the last century has had the effect of decreasing the collective strength of all. Wisdom would seem to indicate that the family of High Energy Physics must somehow learn to resolve its differences and speak with a unified voice.'

The rationale of the new strategy was expressed by Lee DuBridge before the NAS: 'we know our own field is of great importance and we all know that our own field is grossly underfunded. Often we may be tempted to argue that certain fields are overfunded. I hope this temptation can be avoided, at least in our public statements. Our objective should be to increase the total support of basic science.' Since the specialized professional societies are organized around specific fields, it was only natural that the initiative for the strategy of restrained competition should originate in the comprehensive scientific institutions (NAS, NSF and PSAC), and among former Presidential advisers, all of whom enjoy a synoptical view of science as a whole as well as a profound knowledge of its political condition.[12] The new rationale

12. This does not necessarily mean that these scientist-statesmen are easily recognizable by their political skills. Often the level of their political sophistication appears to be correlated with the degree in which they preserve the appearance of political innocence. They know very well that the authority and influence of scientists in politics largely depends, as Robert Wood once pointed out, on their ability to appear apolitical. They also know to distinguish between the political uses of 'political innocence' and the political costs of political insensitivity.

has increasingly acquired the status of a collective political consciousness. Its central idea is that in the long run each scientific field will be better off if most of the scientific community will endorse (or at least not criticize) the claims of some fields at given times than if all the sciences were to present their maximal claims all the time. The need to co-ordinate the strategy of each field with that of all the rest has naturally implied a growing role for such scientific bodies as the NAS, NSF and AAAS, which are capable of articulating the collective authority of science as against the particularistic and more parochial pronouncements of the specialized scientific associations. In this way, some of the authority and influence that such multi-disciplinary organizations of science had lost in the past, because of the fragmentation of the scientific community, seems to have been recovered. While their comprehensive overview of science has made these organizations particularly useful to the Congress and the Executive, their public status as spokesmen for science as a whole has in turn strengthened them *vis-à-vis* the specialized scientific associations and societies. This trend has, of course, the healthy effect of separating central political functions from the professional scientific functions of specialized societies, and has forced the individual scientific fields to build up considerable support among scientists of other fields before presenting their case in the open political arena. The NAS has devised COSPUP to be a central reference point in this process, both as a co-ordinator and a buffer between science and politics.

The NAS–COSPUP has set out to strengthen the political resources of those scientific fields whose intellectual merit is not matched by their political prowess, and to economize the political resources of science by designing selective and well-prepared exposure of various scientific fields. Paradoxically, the actions of COSPUP have amounted to the use of political techniques to protect the traditional autonomy of scientific norms from external political pressures. It reflects the growing consciousness within the scientific community of the political condition of science, of the relationships between the popular images of science, and of the welfare of the scientific enterprise. The Academy, together with other comprehensive scientific organizations and many informal groups of scientific influentials, has attempted to improve the positive political visibility of scientific fields by linking them to areas of public concern and popular technologies, by increasing their accessibility to the public, and by encouraging consensus in public forums. The astronomers, for instance, like the high energy physicists, were encouraged to close ranks within their respective communities before stating their case to the political authorities; the mathematical community and other scientific fields have been helped by NAS–COSPUP in producing field reports to describe their objectives and needs in common language; the physical sciences have been quietly helped to offset links with unpopular military technology by giving greater attention to social problems such as pollution; and fields like genetics have been backed in their efforts to withhold unqualified sanction to genetic facts or theories with highly explosive political connotations.

It is still too early to evaluate this new political orientation of American scientists, and the extent to which their responsiveness to external political opportunities and demands is consistent, in the long run, with the preservation of the internal sub-culture of science. But if growth is to be a measure of success, and the spectacular history of American science since the Second World War is to be its testimony, it would seem that it is no longer the political asceticism of scientists, but rather their conscious, adaptable and economic utilization of their political resources, which will best serve the advancement of science.

References

BAKER, J. J. W., and ALLEN, G. E. (1968), *Hypothesis, Prediction and Implication in Biology*, Addison-Wesley.

BODMER, W. F., and CAVALLI-SFORZA, L. L. (1970), 'Intelligence and Race: it seems fruitless to enquire if differences in IQ have a genetic basis', *Sci. Am.*, vol. 223, no. 3 (October), pp. 19–29.

BUTTERFIELD, H. (1966), *The Origins of Modern Science*, New York, rev. edn.

Congressional Records (1969a), E.6844, 12 August.

Congressional Records (1969b), E.9348, 5 November.

Congressional Records (1969c), E.10910, 20 December.

EZRAHI, Y. (forthcoming), *Public Policy*.

GILLISPIE, C. C. (1960), *The Edge of Objectivity*, Princeton.

JENSEN, A. (1969), 'How much can we boost IQ and scholastic achievement?', *Evnironment, Heredity and Intelligence*, Harvard Education Review, Reprint Series, no. 2, (June).

JENSEN, A., LEWONTIN, R. C., and RABINOVITCH, J. (1970b), (Separate contributions), *Bull. At. Sci.*, 26, no. 5, May, pp. 17–26.
LEWONTIN, R. C. (1970a), 'Races and intelligence', *Bull. At. Sci.*, 26, no. 3, March, pp. 2–8.
LIFE (1970), 'A scientist's variations of a disturbing racial theme', *Life*, 12 June.
MATHER, J. C., and NEURATH, P. (1959), 'An Indian experiment in farm radio forums', UNESCO Series, pp. 61–111.
MCRAE, R. (1961), *The Problem of the Unity of the Sciences: Bacon to Kant*, Toronto.
MILLER, P. (1965), *The Life of the Mind in America*, Harcourt, Brace & World.
NEW YORK TIMES (1969), 'Jensenism, the theory that IQ is largely determined by the genes', *New York Times Magazine*, 21 September.
PRICE, D. K. (1962), *Government and Science*, Oxford University Press.
PRICE, D. K. (1963), 'Science and the race problem', *Science*, no. 142, 1 November.
PRICE, D. K. (1965), *The Scientific Estate*, Harvard University Press.
Science Policy Research Division, Legislative Reference Service, Library of Congress (1969), 'Technical information for congress: report to the sub-committee on science, research and development of the committee of science and astronantics, US house of representatives 91st congress, first session', Washington.
TOCQUEVILLE, A. de (1945), *Democracy in America*, New York.
TURNER, J. S. (1970), *The Chemical Feast*, New York.

13 Paul Gary Werskey

British Scientists and 'Outsider' Politics 1931–1945

P. G. Werskey, 'British scientists and "outsider" politics 1931–1945', *Science Studies*, vol. 1, no. 1, 1971.

A slightly different version of this paper was presented on 4 September 1970 in Durham to a joint meeting of Sections N and X of the British Association for the Advancement of Science. The author would like to thank Dr C. P. Blacker, Professors A. V. Hill and Lancelot Hogben, Sir Julian Huxley and Dr W. A. Wooster for their assistance in his research.

Much has been written in recent years about the political activities of natural scientists in Great Britain during the 1930s.[1] For the most part, such literature has concerned itself with certain ideological affinities which encouraged an alliance between 'moderate' and 'left-wing' elements within the scientific community. In this paper, however, emphasis will be placed on important ideological divisions between the leaders of the Science and Society Movement. The fact that a scientists' 'popular front' was achieved will be explained in terms both of the low status accorded their profession by the nation's political and intellectual élites up until 1939, and of the subsequent demand for scientific expertise during the Second World War. As political 'outsiders', it was therefore natural that at least some British scientists attempted to influence public policy through the formation of pressure groups independent of both the Government and party politics.[2] After discussing the development and decline of the alliance of socially conscious researchers, the paper will conclude

1. The most general survey of the area is provided in Wood (1959, especially pp. 121–51), but note also King (1968, pp. 34–73). Still useful are: Barber (1952), and Crowther (1941, pp. 600–32). For brief comments which rely upon Wood, see Coser (1966, pp. 233–41) and Vig (1968, especially pp. 25–7). See also Rose and Rose (1969, pp. 51–7).

2. Compare Gay (1969). It should also be noted that, under certain conditions, political 'insiders' may find it useful to engage in 'outsider' politics.

with a brief comparison between the scientists' movements of the Thirties and those of the present day.

Scientists as political 'outsiders'

British scientists came to politics during the 1930s as 'outsiders' in two ways. In the first place, civil servants and party politicians tended to exclude men of science from high-level Government appointments. The institutional discrimination against scientific workers within the Civil Service was a much discussed subject at the time. (See for example, Menzler, 1929; Gardiner, 1931; Gregory, 1936; Hutchinson, 1970.) Of greater significance, however, was the evident lack of interest shown by most professional politicians in the social ramifications of scientific research. Thus Solly Zuckerman could argue (1939) 'the efforts of scientists are generally misunderstood, because they are not interpreted to the world by scientists themselves, and because few of those who are immediately responsible for the conduct of social affairs are scientists. There are, for example, no scientists in the Government.'

And Zuckerman could have gone further. With the exception of Sir Oswald Mosley, no British politician in the 1930s was to be heard arguing the case for science as forcibly as had, for example, Balfour and Haldane during the preceding decade. Mosley, on the other hand, could not attract more than a handful of researchers into his British Union of Fascists, even though he described his ideology as a blend of 'Caesarism and Science'.[3] Given that his movement was closely identified with those continental régimes which abused natural science, it is not surprising that Mosley found few adherents among British scientists. In retrospect it is clear that political influence was not denied to all researchers.[4] None the less, it was widely believed at the time that, apart from matters of science policy, the channels of conventional politics were closed to natural scientists.

3. Mosley (1968, especially pp. 316–35). The only example encountered in my research of a scientist who was a member of the British Union of Fascists was Capt. George Pitt-Rivers, one-time President of the International Organization of Eugenists. See Keith (1950, pp. 552–3).

4. Compare Snow (1962), for the cases of Lindemann and Tizard. See also the study by MacLeod (forthcoming). This group of political 'insiders' was largely responsible for the shape and direction of the Government-supported scientific research during the inter-war period.

Scientists, of course, had the option of becoming academic dissenters, a role sometimes assumed by members of Britain's intellectual élite. But here, too, many scientists, precisely because of their backgrounds, found themselves regarded by artists and social theorists as outsiders, merely to be tolerated. The anti-scientist phenomenon is particularly interesting on the political left, where 'Science' was highly esteemed. (Note for example, Strachey, 1933, 117–18; Upward, 1969; Webb and Webb, 1937, pp. 1132–4.) Strachey (1932, p. 177), for instance, chose to characterize scientific men as, 'for the most part, rather simple minded fellows outside of their laboratories'. He then went on to dissect a leading article by Linn Cass (1930) which symbolized to him 'the growing loss of self-esteem and self-confidence which they [the scientists] are feeling, and which they must continue progressively to feel, in the capitalist world'. (Strachey, 1932, p. 180.) After ridiculing the suggestion of the editorial's author that a reversion to 'cottage industries' might resolve Britain's economic diffities, Strachey concluded: 'The hope which he brings to the unemployed is the hope of the destruction of science. If he has to choose between capitalism and science, he chooses capitalism every time. For he is a spokesman of the capitalist class, long before he is a scientist.'[5] An even more savage attack upon natural scientists was presented by Gorer (1936) in an important novel of the mid-1930s. Gorer (pp. 118–19) confronts his hero (Freddy Green) with one Roger Hairwate, a geneticist who speaks about some poison gas research being conducted at his university:

'We do a lot of it. Government grant you know. Of course the research is nominally on insecticides.'

'But it's monstrous! Do you mean to say you all put up with it?'

'My dear Green,' said Hairwate, 'we are skilled workers, not dreamers. Our business is to find out as much as we can about certain phenomena. What these phenomena are to be is decided partly by us and partly by the people who pay us. The use that is made of our

5. King (1968, p. 56) has taken exception to Strachey's view: 'Any loyal reader of the *Nature* editorials for the post-war years would have known that these strictures were not altogether just.' King's general dictum does not, however, apply to the leader in question. Further, given the retreat of the journal in 1938 from its tentative acceptance of 'planning' in the early thirties, one is compelled to admire the prescience of Strachey's polemic. See Werskey (1969).

research is not our business. Our business is exclusively with chemical and biological facts. I may personally deplore the uses to which some discoveries are put, but that is neither here nor there. My business is to do my job.'

Examples of this attitude can be multiplied. Thus the young poet Bell maintained in 1928 at the Cambridge Union (on the motion 'That the Sciences are murdering the Arts') that 'The scientist, the inquirer, the interrogator, was innately incapable of either creating or appreciating art. The business man, the waste product of Science, was the immediate murderer.'[6] At Oxford, Spender formed the opinion that scientific workers were, generally speaking, crude, insensitive bores.

> ... they had a passion for women, combined with an almost complete inability to understand them. ... They researched into life, pretending that their behaviour was an inquiry yielding results, like experiments in laboratories. ... The scientists talked about music with an air of complete familiarity with the scores, and with the lives of the composers. ... The composers seemed to them much like themselves (as perhaps indeed they were), a technical, clever, virile, beer-drinking and coarse-mouthed race (1953, p. 41).

In short, the respective guardians of Britain's political and cultural welfare in the 1930s had little use for scientists.

This conjunction of political and intellectual conventions opposed to scientists came at a time of acute domestic and international crises, represented by the familiar litany of depression, fascism and war. Some of these developments directly concerned the scientific community. The Hunger Marchers in Britain symbolized the menace of malnutrition which Boyd Orr (1936) later quantified in his study of the relationship between food, health and income. From Germany came the anti-semitic nightmare of university purges, compulsory sterilization laws and 'Nordic' racial theories often directed at scientists of great renown. (See Brady, 1937, especially pp. 39–77.) Finally there was an increasingly insistent demand for scientific assistance in building up the nation's defences. For the minority of natural scientists who were politically conscious – and they were a very small minority indeed – such issues could not be ignored. The fact that their newly aroused social concern made very little difference to either political 'in-

6. As quoted in Stansky and Abrahams (1966, p. 46).

siders' or academic outsiders helped to unite (for a time researchers of very different political persuasions.[7] In pointing to their unanimity on the question of raising the social status of scientists, subsequent commentators have in fact spoken of a '"social relations of science movement" that seemed almost to dominate the British scientific world between 1932 and 1945'. (Wood, 1959, p. 121.)

Reformists and radicals

On close inspection, this 'movement' appears to have been neither monolithic nor cohesive. In the first place it involved no more than a few members of the scientific community. Researchers employed by industry or the Government were conspicuously absent. Moreover, the alliance between socially conscious academic scientists was at best a tenuous one. Permanently divided between what might be called its Reformist and Radical factions, the movement could only have survived in a political atmosphere which enforced upon scientists a sense of national unity and emphasized their long-standing professional grievances. This atmosphere prevailed just before and during the Second World War. After the war, in a period of less pressing circumstances, ideological differences became more visible, personalities achieved greater prominence, and the movement divided and gradually faded away.

The basic dissimilarities between the Reformists and Radicals of the 1930s can be easily summarized. The former were for the most part prepared to accept the social order as it was, provided that they and their kind were given a greater voice in public affairs.[8] The fact that the Reformists saw themselves as ostracized

7. Another factor making for unity among the scientists was the Government's cutback in resources available for scientific research during the Depression years. Even though such 'retrenchment' began to ease in 1933, the belief persisted among politically active scientists that British science suffered from insufficient financial support. See Huxley (1934), and Bernal (1939).

8. Werskey (1969, pp. 468–72). The Reformists consisted of senior scientists, some of whom were experienced political 'insiders'. Sir Daniel Hall, for example, had occupied a key position in the Ministry of Agriculture during the first Labour Government. He later served as chairman of the Development Commission. See MacLeod (forthcoming). The unspoken leader of the 'reformists' was Sir Richard Gregory, the editor of *Nature*. See Armytage (1957). Other important Reformists were Sir Frederick Gowland Hopkins, Julian Huxley, Sir John Boyd Orr and Lord Stamp.

from a political system they ultimately supported was an anomaly which they attempted to resolve during the mid-1930s. After an earlier flirtation with a variety of planning doctrines, the Reformists came to embrace more orthodox views of politics and economics. The Radicals, on the other hand, believed that only a society transformed along socialist lines would be prepared to make the fullest and most humane use of scientists and their discoveries.[9] They presented their plea for an improvement in the cultural and political status of the scientist as an essential but subsidiary clause in their demand for a broad social revolution.

Given such a divergence of world-views, it is not surprising that before 1938 these two groups tended to work through different kinds of organization. The Reformists concentrated most of their attention on the British Association for the Advancement of Science. Their ginger group, prior to its assimilation into the BA in 1936, was the British Science Guild. Both organizations were assured of a sympathetic and influential platform in *Nature*, whose editor, Sir Richard Gregory, was the chief spokesman for the Reformists cause. The Radicals, on the other hand, were involved with the activities of not only the Association of Scientific Workers and the Cambridge Scientists' Anti-War Group,[10] but the Labour and Communist Parties too. Their writings appeared in both the ASW's *Scientific Worker* and the Marxist *Modern Quarterly*.

If the theories and practices of the two groups were so dissimilar, why have they so often been bracketed together? One factor which blurred the distinction between Reformists and Radicals was the similarity of their rhetoric on certain key issues. One of these was the theme of 'social responsibility in science'. The Reformists also came to accept the Radicals' stand on both eugenics and the

9. Statements about the Radicals contained in this paper are for the most part based on the author's as yet uncompleted doctoral dissertation: 'The Visible College: A Study of Radical Scientists in Britain, 1918–1939'. The principal Radicals were J. D. Bernal, P. M. S. Blackett, J. B. S. Haldane, Lancelot Hogben, Hyman Levy, Joseph Needham, C. H. Waddington and W. A. Wooster.

10. Nothing has as yet been published on the Radical takeover of the ASW; see E. K. Andrews's forthcoming study of the association. Information derived from the author's conversations with the then Hon. General Secretary, Dr W. A. Wooster, will be included in 'The Visible College' (the author's as yet uncompleted doctoral dissertation). On the Cambridge group, see Burhop (1966, pp. 32–42).

'social relations of science'. Between 1924 and 1936, editorials in *Nature* had echoed the anxieties of the Eugenics Society about so-called 'racial decay'. The journal had advocated, among other things, compulsory sterilization for the unemployed on the assumption that indigence was a sign of sub-normal intelligence. (See Werskey, 1969, pp. 467–8.) Of course, the Radicals had long argued against such a position by pointing to the way in which environmental circumstances helped to determine an individual's place in society, irrespective of his abilities. (See Haldane, 1932.) Until 1934, the Reformists had also stressed the impact of science *on* society,[11] while Radicals spoke in terms of an interaction *between* science and society. But between 1934 and 1936, the Reformists altered their approach to the social relations of science; they had, meanwhile, also changed their attitudes to eugenics.

In both instances the agent and symbol of ideological change was Julian (now Sir Julian) Huxley. The eugenics issue was a case in point. Before 1935 Huxley had been an enthusiastic eugenist, asserting that the inherited potentialities of slum dwellers were below average. This was 'almost certainly not due to the effect of living generation after generation in the slums, but to the fact that a considerable proportion of types that have inherited poor qualities have gradually drifted into slum conditions of living'. (Huxley, 1926, p. 41.) Huxley had warned of the tendency 'for the stupid to inherit the earth, and the shiftless, and the imprudent, and the dull. And this is a prospect neither scriptural nor attractive'. (1931, p. 109.) As a measure to hold down the birth rate of the working classes Huxley advocated (during the worst part of the economic slump) that the continuance of unemployment relief be made conditional upon a man's having no more children. 'Infringement of this order could probably be met by a short period of segregation in a labour camp. After three or six months' separation from his wife he would be likely to be more careful the next time.' (1931, p. 88.) Huxley's views were comparable to the Reformists' approach to eugenics, as expressed in the leaders of *Nature* before 1936.

11. The Reformists' stand in this area has been dealt with in Paul Gary Werskey, 'Planning and Professionalism: *Nature* on the Organization of Science between the Wars' (typescript, Science Studies Unit, Edinburgh University). This essay will shortly appear in *Nature*.

In the early 1930s, however, Huxley's friend, Lancelot Hogben, had begun to mount a fierce attack upon the social Darwinism latent in the thinking of most eugenists. (See Hogben, 1930, pp. 193–215; 1931; 1933.) Hogben's campaign, which labelled the orthodox eugenist as a neo-Nazi, made a profound impact upon Huxley.[12] The resulting shift in Huxley's views on eugenic practice could be seen in his 1936 Galton Lecture, 'Eugenics and Society'. (Reprinted in Huxley, 1941, pp. 34–84.) After quoting Hogben on the need for equalizing educational opportunity before measuring levels of intelligence, he went on to argue that 'we shall only progress in our attempt to disentangle the effects of nature from those of nurture in so far as we follow the footsteps of the geneticist and equalize environment.... We must therefore concentrate on producing a single equalized environment. ...' (1941, p. 69.) Shortly thereafter, *Nature* featured Huxley's lecture in the editorial which signalled the Reformists' retreat from the eugenic approach to social problems. (Crew, 1936.)

On the social relations of science, Huxley's thinking was shaped significantly by the Marxist mathematician Hyman Levy. In 1933 Huxley gave a series of talks on the BBC entitled 'Scientific Research and Social Needs', which were published a year later.[13] Discussions between Huxley and Levy opened and closed the series. At the start Levy asked his biologist companion how he would define science. The reply was:

> Well, I generally like to think of science as a body of knowledge. ... This knowledge can generally be applied to controlling nature, but most scientists, I think, would say that there definitely is something that can be called *pure science*, which has a momentum of its own and goes on growing irrespective of its applications. (1934, pp. 15–16.)

To which Levy replied: 'Well, Huxley, I think that to state things

12. This is borne out by Dr C. P. Blacker, then General Secretary of the Eugenics Society, in an interview with the author, 7 August 1969. This was also Hogben's view, as confirmed in an interview with the author, 26 July 1968.

13. In his autobiographical *Memories*, Huxley (1970) refers to a book called *Science and Social Needs* (written with the assistance of J. G. Crowther). I have not come across a book with that title; presumably he is referring to *Scientific Research and Social Needs* (London, 1934). See Werskey (1971).

in this way is to lay a false emphasis on pure science.' (p. 16.) 'It does not seem to me', Levy continued,

that science becomes 'pure' because there are individual scientific workers whose personal motive in carrying through investigations is that they desire simply to extend the boundaries of knowledge. The existence of such a motive does not necessarily enable them to lift themselves outside their historic social epoch, but it may mean that they will concentrate their attention on problems more remote from direct application. Science, however, does not cease at discovery. It is also concerned with application, and the applications are to the systems of society in being.... Moreover, since scientists, like other workers, have to earn their living,... to a large extent the demands of those who provide the money will, very broadly, determine the spread of scientific interest in the field of applied science.... I know of no scientist who is so free that he can study absolutely anything he likes, or who is not restricted in some way by limitations such as the cost of equipment. (p. 20.)

After this interchange Huxley conducted a tour of different research establishments. By the final broadcast he was persuaded that 'the form and direction which it [science] takes are largely determined by the social and economic needs of the place and period'. (p. 252.) Huxley concluded that the chief moral of the series was that

science is not the disembodied sort of activity that some people would make out, engaged on the abstract task of pursuing universal truth, but a social function intimately linked up with human history and human destiny. And the sooner scientists as a body realize this and organize their activities on that basis, the better both for science and for society. (p. 279.)

The Reformists took Huxley's moral to heart. Shortly after the publication of his broadcasts, Brightman (1934), the chief leader writer for *Nature*, proclaimed that:

The conception of science as a social function intimately linked up with human history and human destiny, moulding and being moulded by social forces, should summon forth from scientific workers something of the energy required to translate into policy and action the knowledge acquired by their work. Such energy will find its expression... in... the faith that human reason, by using wisely the scientific method, can give us the control of our destiny.

From that point onwards the Reformists began to employ the

vocabulary of the Radicals when speaking about the social relations of science.

It is difficult to judge whether the pronouncements of Huxley were in themselves an important factor in the determination of Reformist thinking, or whether Huxley's writings just happened to reinforce conclusions which the Reformists had reached independently.[14] In either case his rhetoric permeated some of the most crucial leading articles which appeared in *Nature* during the mid-1930s. Thus the Radicals, through the modifications they introduced into Huxley's social thought, were able to influence somewhat the world-view of the Reformists. It is also evident that Huxley's apolitical stance helped to legitimize the position of his left-wing companions.[15] The fact that the Reformists were never again to take up their old position on eugenics must be counted as a victory for the Radicals.

'Reformists' *versus* 'Radicals'

But the arguments which supported the notion that social forces affected the course of scientific development were more ambivalent politically than those associated with eugenics. That is why the agreement reached between Reformists and Radicals on the concept of the social relations of science quickly led to an intensification of the ideological divisions between them. To understand how much they differed in practice, one has only to compare their respective responses to the plight of scientists in Nazi Germany and the Soviet Union.

The Radicals argued that, since the scientific community could never fully insulate itself from social pressures, its only course was to align itself with those political forces which were most committed to the advancement of science for the benefit of the entire society. From that premise they maintained that German scientists had only themselves to blame for their situation under Hitler. If they, as university teachers, had not been so apolitical before 1933, they might, the Radicals asserted, have been able to fore-

14. The latter was probably the case as far as eugenics was concerned. See Werskey (1969, pp. 467–8).

15. Huxley lays great stress on the difference between himself and such 'left-wing socialists' as 'Levy and Needham', in a letter to the author, 20 August 1969.

stall the Nazis' rise to power. (See Needham, 1941.) Russian scientists by contrast were in certain respects better off than their counterparts in Britain, or so the Radicals believed. Bernal, Haldane and others repeatedly emphasized the superiority of Soviet scientific organization, the scientific ethos of Russia's leaders, and the comparatively high status accorded scientists in Russian society. Above all, they stressed the way in which scientific resources were devoted to the solution of important economic and social problems. (See, for example, Needham and Davies, 1942.)

But while the Radicals were hopeful about the prospects of a fruitful interaction between science and society, the Reformists were predisposed to pessimism about the effects of systematic social controls on their profession. They were accordingly horrified by the relatively complete integration of scientists into the differing political systems of Germany and the Soviet Union. They believed that the fervent nationalism which informed scientists' attitudes in the two countries was opposed to the values of an international scientific community. From the first the Reformists were opposed to Hitler's Germany on political grounds (see Fallaize, 1933), but when scientists loyal to the Führer also attempted 'to secure the control of international scientific work', a *Nature* editorial urged that 'it is time to call a halt'. (See Stratton, 1935.) Russia was subjected to similar criticism. (See Marvin, 1932.) The Reformists also rejected the Fascists and the Communists for consciously curtailing the intellectual freedom of individual scientists. In this respect, the Reformists were more concerned about Germany than Russia (see Adams, 1937); but they also used Lysenko's early harassment of well-known orthodox geneticists to illustrate 'the atmosphere in which scientific investigators in totalitarian countries have to live and work'. (See Uvarov, 1937.) In equating the threats posed by Hitler and Stalin to the internationalism and freedom of science, the Reformists were prompted to support the exclusion of both Germany and Russia from the councils of world science 'on the ground that in these countries at present scientific workers are bound much more closely to their respective Governments than is the case elsewhere'. (See Brightman, 1937.)

The disparity between the Reformist and Radical scientists was therefore never greater than it was in the summer of 1938. As the Reformists' awareness 'of the extent to which political organizations can affect the direction of scientific research, and even frustrate its efforts' increased (see Brightman, 1936), their interest in the extension of state responsibility for science diminished. At the same time Bernal was completing his *The Social Function of Science* – the Radicals' most comprehensive blueprint for the reorganization of scientific life. (1939, especially pp. 241–416.)

The scientists' popular front

Paradoxically, in the midst of such intense political disagreements, the Reformist and Radical factions came together in 1938 to found the Division for the Social and International Relations of Science in the British Association.[16] The origins and development of this new organization are therefore of considerable interest.

The idea for such a division originated in the Committee on Science and its Social Relations set up (in 1937) by the International Council of Scientific Unions. (Stratton, 1937 and also 'Committee on social contacts of science', *Nature*, 1937.) The President and Vice-President of the committee were, respectively, F. J. M. Stratton and Sidney Chapman, both of whom happened to be regular contributors to *Nature*. When the Council of the Royal Society refused to establish a working group in this area,[17] Chapman and Stratton went to the Council of the British Association, which by this time was controlled by the Reformists. There it was decided that the kind of problems related to the social relations of science would best be handled by a division set apart from the parent body. The advantages of this format would include more frequent meetings and independent publication. Most important, no corporate decisions would be taken by the new organization on matters of social policy. Instead the 'purpose of the Division', as explained at the time by the Assistant Editor of *Nature*, 'would be to further the objective study of the social relations of science. The problems with which it would deal would be concerned with the effects of the advances in science on the well-being of the community, and, reciprocally, the effects of

16. See the supplement 'Social Relations of Science', *Nature* (1938).

17. Personal communication, Professor A. V. Hill (then a member of the Royal Society's Council) to author, 9 May 1969.

social conditions upon advances in science.' (See Gale, 1938.) The role of the new Division was *discussion*; it was carefully designed *not* to become the kind of forum proposed by Reformists in previous years and later realized by the Radicals in the Association of Scientific Workers – a forum from which a 'united front' of scientists could speak out on controversial questions.[18]

The reason for the Council's reluctance to provoke controversy was not far to seek. As one of *Nature*'s correspondents had commented in 1932, science 'has never gotten within sight of a political programme of its own. . . .' (See Fenn, 1932.) By 1938 the validity of that statement was only too clear, even within the 'leader' columns of *Nature*. Thus Tripp (1937), a year before the establishment of the BA's new division, had confidently predicted that

> The future historian of science will not fail to chronicle that the early part of the twentieth century was notable for the gradual emergence of a social conscience among scientific men, which, he will aver, was greatly stimulated by the mis-use of certain scientific discoveries for inhuman ends (for example, poison gas against civilians), and by the recognition that extending [the] application of science to industry did not appreciably improve the status or the prospects of the working classes.

Subsequent historians have appeared to take Tripp seriously, and his description is, in general terms, apt when applied to the Radicals.

Yet his reasoning had already become inapplicable to the Reformists by the time of the British Association's Cambridge meeting of 1938. The content of their social concern, as reflected in the dramatic shifts in *Nature*'s leader policy, was now defined not only or even principally in terms of social welfare but in terms of the freedom of science as well. Soon after the BA had launched its new organization, a *Nature* editorial maintained that

> The widespread interest in the social relations to [*sic*] science . . . has largely been stimulated by the growing anarchy in the international sphere, whether economic or political. The threat to freedom of thought inherent in the Totalitarian States, the existence of which is indeed only possible through the application of scientific knowledge, provides one of the main stimulants. The profound concern engendered everywhere by the increasing scale . . . [of] preparations for warfare, even to the

18. For the Reformists' earlier ideas in this area see Campbell (1923) and Brightman (1936).

detriment of standards of living, however, provides another source of such interest. . . . (Brightman, 1938.)

Obviously the sources of social awareness varied among the activist scientists according to their ideological commitments. Within a few years, in fact, a fierce internecine battle broke out within the scientific community between those who stressed the social function of science and those who emphasized the need for freedom in science. Nevertheless, Reformists and Radicals were able to work with one another up until 1945, not only in the BA's new division,[19] but also in Zuckerman's celebrated Tots and Quots Club.[20]

This alliance was welded by two related factors. First, both Reformist and Radical factions were anxious about what might happen to Britain if it continued to treat its scientists as political outsiders. Their despair only deepened when Britain entered the Second World War.[21] Undoubtedly the activist scientists were able to keep together, because a British victory appeared to depend greatly on the effective utilization of scientific expertise. The fact that *Nature*'s first detailed proposal for the mobilization of scientists in wartime was (anonymously) written by Bernal (1938) indicates how this sense of national crisis worked to bring the rival wings together.

The demise of the popular front

But the 'popular front' created by the two factions could not outlive the pressures of war. As early as 1941 a Society for Freedom in Science had been formed by Baker and Polanyi, among others, to provide a 'liberal' alternative to the Radical position. (See Wood, 1959, pp. 134–6. Also, the 'Occasional Papers' of the Society for Freedom in Science between 1940 and 1960; Baker, 1942, 1945; Polanyi, 1951.) Baker and Polanyi occasionally at-

19. See Crowther, Howarth and Riley (1942), for an account of the Division's Science and World Order Conference of September 1941. Gregory, who was President of the BA throughout the war years, organized a number of meetings of this kind.

20. The Tots and Quots meetings deserve an article to themselves. In the meantime see Zuckerman (1967, pp. 147–8), and Crowther (1970, pp. 210–22).

21. Note the 'Penguin Special' prepared by the Tots and Quots Club: *Science in War* (1940).

tacked the Reformists as well, but the latter were for the most part already persuaded of the necessity for preserving at all costs the operational autonomy of the scientific community. For that matter, so were the Radicals, at least in theory; the plan advocated by Bernal in the second half of *The Social Function of Science*, for instance, would in certain crucial respects have increased the independence of the scientific community from its social patrons. (See especially pp. 310–22.)

The Radicals were pilloried both during and after the war as the enemies of freedom in science for two reasons. First, the Radicals believed that their proposals for the reorganization of scientific life were already being carried out in the Soviet Union. Although this was not true, Baker, Polanyi and others quite willingly took the Radicals at their word and cited the damaging example of Stalin's interference with the Russian scientific community as a warning against the ideas of Bernal and his associates.[22] Thus the Radicals paid dearly for their often uncritical admiration of Soviet Communism.[23] Second, as Marxists, the Radicals had long derided the social and philosophical connotations of the concept of 'pure science'. They were therefore viewed by the Society for Freedom in Science as individuals who sought the destruction of not merely the concept but the practice of pure science, leaving only armies of technologists to carry on the scientific tradition. (See Baker, 1939.) In the disturbed political atmosphere of the early post-war years, the fact that the Radicals recognized the continuing need for 'fundamental research' was overlooked.[24]

What finally split the Radicals and the Reformists, however, was not the activities of the Society for Freedom in Science, but a series of domestic political developments over which neither group had control. The most important of these was a gradual improvement in the Government's treatment of science and scientists – expenditure for research greatly increased and scientific men were seen more frequently in the corridors of power.

22. The qualified apologies of Bernal and Haldane for Lysenkoism did not, to say the least, strengthen their position in this area. See Haldane (1948); and Bernal (1954, pp. 665–73).

23. As had been predicted by Waldemar Kaempffert. See his review of Bernal's *The Social Function of Science*, (Kaempffert, 1939).

24. Note Bernal's reply (1939) to Baker.

(Vig, 1968.) Thus the principal planks of the Reformists' old platform had been adopted.[25] Having been granted a certain amount of political and financial recognition, many activist scientists concluded that their crusading days were over. The British Association simply allowed its Division for the Social and International Relations of Science to wither away during the Fifties. What spirit of activism remained was channelled into the World Federation of Scientific Workers, 'Science for Peace', and the Campaign for Nuclear Disarmament.

Outsider politics, 1971

Given certain recent developments,[26] it may be appropriate to conclude this essay with some speculative comparisons between the scientific organizations of the interwar period and the new British Society for Social Responsibility in Science. (See Rose and Rose, 1969.)

Some scientists now feel themselves obliged to take up defensive positions and to resort again to the concerns of 'outsider' politics. During the mid and late 1960s the successes of science and technology were themselves called into question, if only because they appeared to threaten the existence of 'post-industrial' society: for example, new advances in chemical, biological and nuclear warfare; higher levels of pollution; and the unforeseen social and economic consequences of pursuing industrial automation for its own sake.[27]

In Britain, the British Society for Social Responsibility in Science has been among the most vocal groups expressing this concern. But, at least at this stage of its development, the BSSRS has neither the relatively broad-based support once enjoyed by the Division for the Social and International Relations of Science, nor the ideological bite of the old ASW. The fact that this new

25. Although scientists were largely called upon to advise on, rather than make, Government policy: Crowther (1967), pp. 119–25.

26. Namely the activities of a ginger group, supported by the British Society for Social Responsibility in Science, during the recent Durham meeting of the British Association. See Pirani (1970).

27. An excellent monograph summarizing the literature dealing with such problems has been prepared, but not as yet published, by Dr Roger Williams of the University of Manchester. A shortened version was presented at the Durham meeting of the British Association, 4 September 1970.

society has had to perform a variety of tasks previously distributed between Reformist, Radical, and 'popular front' groups has undoubtedly contributed to its blurred image. (See Tucker, 1970.)

Yet even if the BSSRS does succeed eventually in 'getting itself together', it is doubtful whether it (or any similar organization) will be able to persuade the scientific community at large to by-pass, or even supplement, 'insider' channels in favour of other types of action. As in the 1930s, activist scientists are ideologically divided. Unlike the interwar period, however, there now appears to be no overriding professional grievance which might unite large numbers of researchers. More to the point, scientists are now less often excluded from positions of administrative and political influence. Thus one political 'outsider' of thirty years ago, Zuckerman, is presently Chief Scientific Adviser to HM Government. From his Whitehall office, according to the *Medical News Tribune*, Sir Solly now views the recrudescence of political frustration among some scientists in the following terms:

These people concerned with – what's it called? Oh yes, social responsibility in science: they're probably worried because they see science being misapplied.

They probably believe that the people responsible – if they can be identified – are necessarily evil. I don't know whether it is as simple as that.

What they have to ask themselves if they wish to be effective is whether they can influence the tide of events in the application of scientific knowledge, from the outside.

There is a certain . . . naivety in the assumption that you can . . . stay away over there (he gestures towards Horse Guards Parade) shouting this message at some people who are presumed to be somewhere else and doing the wrong thing, and also that they're going to listen. They won't. (Zuckerman, 1970).

If Zuckerman's final prediction is correct, then it is possible that at some point in the near future some dissident scientists will begin to emigrate from their relatively secure professional enclaves, like the BSSRS, into the wider and more problematic world of radical politics.[28] And some may go further than that by opting out of science altogether. (See Corbyn and Wield, 1970.) If such developments do occur, then the 1970s will certainly

28. This seems to be already occurring in the United States. See Perl (1970).

become a time when, to employ a famous phrase from the 1930s, British science once again finds itself 'at the crossroads'.[29]

References

ADAMS, W. (1937), 'Freedom of science and learning', *Nature*, vol. 140, 31 July, p. 170.

ARMYTAGE, W. H. G. (1957), *Sir Richard Gregory, his Life and his Work*, London.

BAKER, J. R. (1939), 'Counter-blast to Bernalism', *New Statesman and Nation*, 29 July, pp. 174–5.

BAKER, J. R. (1942), *The Scientific Life*, London.

BAKER, J. R. (1945), *Science and the Planned State*, London.

BARBER, B. (1952), *Science and the Social Order*, New York.

BERNAL, J. D. (1938), 'Science and the national service', *Nature*, vol. 142, 15 October, pp. 685–7.

BERNAL, J. D. (1939a), 'Reply to Baker's "Counterblast to Bernalism"', *New Statesman and Nation*, 5 August, pp. 210–11.

BERNAL, J. D. (1939b), *The Social Function of Science*, London.

BERNAL, J. D. (1954), *Science in History*, London.

BOYD ORR, J. (1936), *Food, Health and Income*, London.

BRADY, R. (1937), *The Spirit and Structure of German Fascism*, London.

BRIGHTMAN, R. (1934), 'The planning of research', *Nature*, vol. 134, 28 July, p. 119.

BRIGHTMAN, R. (1936a), 'Outlook of professional organizations', *Nature*, vol. 138, 14 November, p. 817.

BRIGHTMAN, R. (1936b), 'The protection of scientific freedom', *Nature*, vol. 137, 13 June, pp. 963–4.

BRIGHTMAN, R. (1937), 'Social responsibilities of science', *Nature*, vol. 139, 24 April, p. 689.

BRIGHTMAN, R. (1938), 'Social and international relations of science', *Nature*, vol. 142, 20 August, p. 310.

BURHOP, E. H. S. (1966), 'Scientists and public affairs', in M. Goldsmith and A. MacKay (eds.), *The Science of Science*, Penguin.

CAMPBELL, N. R. (1923), 'A representative body for science', *Nature*, vol. 112, 13 October, pp. 529–31.

CORBYN, P., and WIELD, D. (1970), 'Science education in a social context', *New Scientist*, vol. 47, 17 September, p. 597.

COSER, L. A. (1966), *Men of Ideas: A Sociologist's View*, New York.

CREW, F. A. E. (1936), *Eugenics and Society*, *Nature*, vol. 137, 11 April, p. 593.

29. *Science at the Crossroads* (Kniga, 1931) brought together the papers presented by the Russian delegation to the International Congress of the History of Science and Technology held in London during the summer of 1931. The book had a profound impact upon the thinking of Radical scientists. This invaluable document, long out of print, will soon be made available as part of the Cass reprint series in the history of science.

CROWTHER, J. G. (1941), *The Social Relations of Science*, New York.
CROWTHER, J. G., HOWARTH, O. J. R., and RILEY, P. P. (1942), *Science and World Order*, Penguin.
CROWTHER, J. G. (1968), *Science in Modern Society*, Cresset Press.
CROWTHER, J. G. (1970), *Fifty Years with Science*, London.
FALLAIZE, E. N. (1933), 'Nationalism and Academic freedom', *Nature*, no. 131, 17 June, pp. 153–5.
FENN, L. A. (1932), 'The politics of science', *Nature*, vol. 129, 19 March, p. 415.
GALE, A. J. V. (1938), 'Social and international relations of science', *Nature*, vol. 142, 27 August, p. 380.
GARDINER, J. A. (1931), 'Scientific men as administrators', *Nature*, vol. 129, 15 August, p. 237.
GAY, P. (1969), Weimar Culture: *The Outsider as Insider*, Secker & Warburg.
GORER, G. (1936), *Nobody Talks Politics*, London.
GREGORY, R. A. (1936), 'The civil service and "everyday" science', *Nature*, vol. 137, 14 March, p. 417.
HALDANE, J. B. S. (1932), *The Inequality of Man and Other Essays*, London.
HALDANE, J. B. S. (1948), 'Lysenko and Darwin', *Daily Worker*, 1 November, p. 2.
HOGBEN, L. (1930), *The Nature of Living Matter*, London.
HOGBEN, L. (1931), *Genetic Principles in Medicine and Social Science*, London.
HOGBEN, L. (1945), *Nature and Nurture*, Fernhill.
HUTCHINSON, E. (1970), 'Scientists as an inferior class', *Minerva*, vol. 8, July 1970, pp. 396–411.
HUXLEY, J. S. (1926), *The Stream of Life*, London.
HUXLEY, J. S. (1931), *What Dare I Think*, London.
HUXLEY, J. S., and CROWTHER, J. G. (1934a), *Scientific Research and Social Needs*, London.
HUXLEY, J. S. (1934b), *Scientific Research and Social Needs*, London.
HUXLEY, J. S. (1941), *The Uniqueness of Man*, London.
HUXLEY, J. S. (1970), *Memoires*, London.
KAEMPFFERT, W. (1939), 'Review of Bernal's *The Social Function of Science*, in the New York Times Book Review, 18 June, pp. 4 et seq.
KEITH, A. (1950), *An Autobiography*, London.
KING, D. (1968), 'Science and the professional dilemma', J. Gould (ed.), *Penguin Social Science Survey*, Penguin.
Kniga Publishing Co. (1931), *Science at the Crossroads*, Kniga.
LINN CASS, W. G. (1930), 'Unemployment and hope', *Nature*, vol. 125, 15 February, p. 225.
MACLEOD, R. (forthcoming), *Application of the Principles of Coordination to Civil and Military Research*, 1919–39.
MARVIN, F. S. (1932), 'Science and society', *Nature*, vol. 129, 5 March, p. 330.

Menzler, F. A. A. (1929), 'The royal commission on the civil service', *Nature*, 12 October, pp. 565–7.
Mosley, O. (1968), *My Life*, London.
Nature Supplement (1928), 'Social relations of science', *Nature*, vol. 141, 23 April, pp. 723–42.
Nature (1937), 'Committee on social contacts of science', *Nature*, vol. 140, 4 December, p. 983.
Needham, J. (1941), *The Nazi Attack on International Science*, London.
Needham, J., and Davies, J. S. (1942), *Science In Soviet Russia*, London.
Penguin Special (1940), *Science in War*, Penguin.
Perl, M. (1970), 'The "new critics" in American science', *New Scientist*, vol. 46, 9 April, pp. 63–5.
Pirani, F. (1970), 'What's wrong with the BA?', *New Scientist*, vol. 47, 3 September, pp. 461–2.
Polanyi, M. (1951), *The Logic of Liberty*, University of Chicago Press.
Rose, H., and Rose, S. (1969a), *Science and Society*, Alan Lane.
Rose, H., and Rose, S. (1969b), 'Knowledge and power', *New Scientist*, vol. 42, 17 April, pp. 108–9.
Snow, C. P. (1962), *Science and Government*, Harvard University Press.
Society For Freedom In Science (1940–1960), 'Occasional papers', Oxford.
Spender, S. (1953), *World Within World*, London.
Stansky, P., and Abrahams, W. (1966), *Journey to the Frontier*, Constable.
Strachey, J. (1933), *The Menace of Fascism*, London.
Stratton, F. J. M. (1935), 'Nazi-Socialism and international science', *Nature*, vol. 136, 14 December, p. 928.
Stratton, F. J. M. (1937), 'International cooperation in Science', *Nature*, 28 August, p. 337.
Tripp, E. H. (1937), 'Science and social responsibility,' *Nature*, vol. 189, 12 June, p. 981.
Tucker, A. (1970), 'Bulls in a scientific talking-shop', Guardian, 2 September, p. 11.
Upward, E. (1969), *In the Thirties*, Penguin.
Uvarov, B. P. (1937), 'Genetics and plant breeding in the USSR', *Nature*, vol. 140, 21 August, p. 297.
Vig, N. J. (1968), *Science and Technology in British Politics*, Pergamon.
Webb, S., and Webb, B. (1937), *Soviet Communism: A New Civilization*, London.
Werskey, P. G. (1969), 'Nature and politics between the wars', *Nature*, vol. 224, 1 November, pp. 462–72.
Werskey, P. G. (1971), 'Haldane and Huxley: the first appraisals', *J. Hist. Biol.*, vol. 4, April, pp. 161–83.
Wood, N. (1959), *Communism and British Intellectuals*, London.
Zuckerman, S. (1939), 'Science and society', *New Statesman and Nation*, 25 February, p. 298.
Zuckerman, S. (1970), *Medical News Tribune*, 27 February, p. 14.

Part Five
Science in Society: Problems of Intelligibility and Impact

Science can have social consequences independent of technology in so far as its concepts and beliefs are absorbed into the general culture. Among the many factors which can influence this process is the dominant image of science. Institutionalised beliefs and stereotypes will always surround science, particularly where it exists in fully differentiated form; among other things they will define its scope and reliability as a knowledge source, and its significance compared with other such sources in the society. Thus they will determine *prima facie* responses to scientific pronouncements and ideas – whether they are worth even listening to, whether they are likely to be true and in what circumstances, what other sources they can be checked against and validated by. In modern society, science enjoys high regard as a knowledge source and the authority of its pronouncements is accepted over a wide area; for many it is the ultimate source for the validation of knowledge, the final court of appeal. It would appear that, as Handlin's paper suggests, these beliefs about science derive from its pragmatic effectiveness as technology, and co-exist in the lay consciousness with elements of distrust and even fear. None the less they ensure that the intellectual products of science get off to a good start in the wider society.

It is noteworthy that Handlin, who shows a genuine curiosity about the credibility of scientific beliefs, is an historian. Sociologists have almost entirely ignored this topic, as indeed they have avoided studying factors influencing the reception of most beliefs regarded as 'knowledge' within their own societies.[1] My own paper tries to identify general principles

1. Bernstein's work is a major exception here; in fact this very point is

guiding the analysis of such questions, and, in particular, points to the importance of intelligibility in the reception of scientific beliefs.

Another important variable here, of course, is whether the beliefs in question effectively attack or support social institutions. When this is so, systematic predispositions must be expected to influence their reception – predispositions influenced by patterns of institutionalised belief, their relationship to institutional structure and subjective commitments to that structure. In a recent discussion of the credibility of the ecology movement Douglas (1970) has pointed out the importance of such factors, enriching her analysis with anthropological material.

Rosenberg's paper is also concerned with the relationship between scientific ideas and social structure, but he is concerned with the concepts and theories of science rather than concrete pronouncements. When these diffuse into the wider culture they link to a different universe of social practice. Rarely does this result in no more than an extension of reference or loss of precision; almost invariably meaning change occurs; theories undergo metaphorical extension and modification as they pass into general use. In such processes, concepts and relationships become remoulded into forms appropriate to actors' socially determined requirements. They come to act, as Rosenberg describes, as metaphors and frameworks for the elucidation and justification of social structure or ideal visions of it.

References

BERNSTEIN, B. (1971), 'On the classification and framing of educational knowledge', in M. Young (ed.), *Knowledge and Control*, Collier Macmillan.

DOUGLAS, M. (1970), 'Environments at risk', *The Times Literary Supplement*, 30 October, pp. 1273–5. (A revised and developed form of this work is published by Longman, 1971.)

---

repeatedly being made within it. Thus, in a recent paper (1971) he writes: 'How a society selects, classifies, distributes, transmits and evaluates the educational knowledge it considers to be public, reflects both the distribution of power and the principles of social control.' Having pointed to the neglect of this topic, he goes on to develop a general sociological theory of the curriculum, essentially Durkheimian in spirit.

14 Oscar Handlin

Ambivalence in the Popular Response to Science

O. Handlin, 'Science and technology in popular culture', in G. Holton (ed.), *Science and Culture*, Houghton Mifflin, 1965.

In our culture, superficial links often obscure the distinction between science and technology. Popular thinking usually blurs the difference; the Cadillac and the space vehicle, the miracle drug and the computer, the H-bomb and dacron are the final products of technology that validate the enterprise of science. Research is development; and what works has a claim to credibility. The fact that two distinct types of activity are conflated in these conjunctions is rarely recognized.

Hence the deep ambivalence in popular attitudes toward science. Rarely has the man in the laboratory been so widely respected; never has he commanded so ready an access to public and private funds. One has only to recall the effects of Sputnik on American education to estimate the value set on his opinions, the esteem accorded his achievements. Yet the people who gladly vote billions for scientific research are far from understanding its inner character; and the points of view associated with it have never been altogether assimilated to the culture even of the West. The 'popular delusions' which the scientist encounters with surprise upon his occasional forays outside the laboratory are the normal beliefs of a world which uses, but does not understand, the learning he develops.

Indeed, a deep underlying distrust of science runs through the accepted attitudes of people in the most advanced nations. Paradoxically, the bubbling retorts, the sparkling wires and the mysterious dials are often regarded as the source of a grave threat. Their white-coated manipulators, in the popular image, have ominously seized a power which they may use to injure mankind.

To disentangle the knotted threads of the relationship between science, technology, and popular attitudes, it is necessary to trace changes that have been almost two centuries in process. The complexities of the present are the product of a long development that has altered each of the elements involved.[1]

We do not have to go very far back in time to locate the prototype of the fixed man in a fixed community who supplies us with a starting point. He occupied the peasant villages of Europe fifty years ago. A century before that, in that continent and in its overseas outposts, he was the predominant figure both in the towns and in the countryside. Survivors still exist today in places that have resisted change.

The characteristic feature of the life of such persons was tradition – an understanding of the universe which passed with minor modification from generation to generation and which anticipated all the decisions the individual was called on to make. The great events of birth, marriage and death, and the lesser ones of sowing and reaping, of digging and building, of contriving and fabricating, were alike governed by a code that was self-validating in that it answered every conceivable question with conviction. The tradition also bore the sanctions of secular and sacred authority, and enjoyed the support of all communal institutions. It marked out an area of action within which man could operate with relatively little freedom but with immense security. The exceptional persons who moved beyond those limits and rejected tradition were individually important for their creativity, but they did not before the nineteenth century greatly influence the mass of men who were content to do and believe as their fathers had.

Tradition governed both the ways of doing and the ways of knowing. It set the patterns by which the artisan guided his tools, the husbandman his plow. It also gave satisfying responses on the occasions when they wondered why; for it supplied a comprehensible explanation for the affairs of the visible and invisible world. The ways of doing were not identical with the ways of knowing but they were associated through common reference points in the traditions of the community.

The disruptive forces that broke in upon these communities already made themselves felt in the eighteenth century; they

1. Some of the themes of this essay were treated in another context in Handlin (1964).

mounted in intensity in the nineteenth; and they have become the dominant factors in the social disorganization of our own time. The migration of vast populations, urbanization and industrialization weakened the cohesive force of inherited institutions. The heterogeneity, impersonality, and individualism of the new social structures sapped the strength of tradition. And novel conditions demanded modes of acting and thinking for which there was no precedent. No matter how earnestly people struggled to preserve or restore the old community and its traditions, they could not stay the transformation.

Their ways of doing and their ways of knowing changed, although in different fashions. In the one case, development was continuous and controlled, in the other abrupt and unpredictable. Technology, however dangerous, had familiar features and bore the promise of service to man; science, however beneficent, was alien and threatened to overwhelm him. Ultimately when the two seemed to fuse, popular understanding failed entirely.

The ways of doing, which became technology, unfolded continuously from the experience of the past. Man the tinkerer had always sought to spare himself labor; and tradition in this respect was not entirely static. The first machines of the eighteenth and the nineteenth centuries were simply extensions of familiar techniques. Social and economic conditions imposed an ever livelier pace on these developments, but they were not altogether novel. There was no sudden severance of the essential continuity of the processes of fabrication. The machines themselves were the inventions of gifted artisans; and their use was assimilated with surprising equanimity.

The experience of factory labor was not altogether discontinuous with the past. The machines, whether of wood or iron, were not totally strange. The waterwheels, the great drive shafts and pulleys that dominated these plants, embodied no essentially new principles. To onlookers they were impressive in their ingenuity and power, but the manner of their operation was clearly visible and seemed but to extend and improve devices with which men had long been acquainted. Nor was the physical setting totally different. The earliest factories appeared in the countryside, not far removed from the familiar landscape of open fields, streams and woods. For a long time they used waterpower, and while the appearance of the mills was more complex than those familiar to every man, essentially they were not too different. It was characteristic, for instance, that one of the earliest American utopian novels to conceive of invention as a way of liberating man from labor had a thoroughly rural setting. In Mary Griffith's *Three Hundred Years Hence* (1836), great machines, moved by some internal power, did all the work of agriculture.

Now and then craftsmen displaced by new contrivances expressed their resentment – in Luddite riots and in political hostility to the growth of corporations. But the masses actually employed at the machines accepted their situation without shock. The village laborer who took a job in the mill was more sensitive to the transition from rural to urban life than to the abandonment of the plow for the loom. Often these people suffered from difficult conditions of labor and from even more difficult conditions of life. But their trials were tempered by a sense of confidence in the human capacity to master the devices that were the products of human ingenuity. They were sure that man could control and use the enormous power of the machines.

Therefore, invention could also be regarded as man's liberator, as Étienne Cabet had suggested it would be in *Voyage en Icarie*. In the series of great expositions that began at the Crystal Palace in London in 1851 and ran down through the end of the nineteenth century, the focal point was often the array of new machinery treated as symbolic of the age. The American commissioners to the Paris Universal Exposition of 1867 published six volumes of observations on the devices exhibited there. In the introduction to their report, Secretary of State William H. Seward explained that it was 'through the universal language of the products of labor' that 'the artisans of all countries hold communication'. Industrialization was 'in the interests of the mass of the people', for it promoted 'an appreciation of the true dignity of labor, and its paramount claims to consideration as the basis of national wealth and power'. Seward was confident that the machine would elevate man to new dignity.

The occasional intellectual observers who expressed a fear of the machine were more likely than not to attach their forebodings to the social situation of the factory. They protested against the bondage of human beings totally controlled by a routine that took no account of personality and detached man from nature. The factory enslaved those who served it by limiting their wills; and since it took no cognizance of moral considerations, it also limited their ability to make choices of good over evil. Here workers assembled in numbers that theretofore had been brought together only through some form of servitude in the military company and the ship's crew, the poorhouse and the gaol – agglomerations in which a rigid discipline that curtailed individual freedom permitted the coordination of many persons. It was no coincidence that the architecture of the early factories had much in common with that of the barracks, the military camp, and the prison. Hence the humanitarian's concern lest the power of the machine in this setting constrict personal liberties.

The factory regime detached work from nature and from all other aspects of life. The machine disregarded the alternations of the seasons and the rising and setting of the sun to operate at its own pace, winter and summer, day and night. There the laborers confronted the enterprise in a relationship that was purely economic. All those who entered the factory did so as detached individuals. Within its gates, they were not members of families or of groups, but isolated integers, each with his own line on the payroll; nothing extraneous counted. During the working hours, the laborers had no other identity than that established by the job. From being people who were parts of households, known by a whole community, they had been reduced to being servants of the machine.

By and large, nevertheless, the nineteenth century clung to its optimism about technology. Edward Bellamy, John Macnie and the utopian novelists of the 1880s and 1890s had no doubt that the machine would liberate mankind through the abundance it created. They did not deny that it would also harness man to its service. But they welcomed the consequent routine, the regularity and the order. Bellamy explained that the idea of *Looking Backward* came to him when he 'recognized in the modern military system' the prototype of the Industrial Army that manned his utopia. The men of the year 2000 had 'simply applied the principle of universal military service . . . to the labor question'. Consequent gains in efficiency and affluence would release energies for the solution of all the problems of freedom raised by industrial regimentation.

H. G. Wells supplied a perfect encyclopedia of these hopes for the future. Beginning with his *Anticipations* (1902), a succession of roseate works showed the machine transforming and improving human life, which would evolve toward ever more centralized control. One state, one language, one ruling will would organize all men into efficient productive units. Indeed there would be no need at all for human labor as a source of energy. 'Were our political and social and moral devices only as well contrived to their ends as a linotype machine . . . or an electric tram-car, there need now . . . be no appreciable toil in the world.' Despite the anticipatory fears of those concerned with the future of man's spirit, in the last analysis there was faith that the machine remained a product of man and would obey his command.

Changes in the ways of knowing aroused far greater uneasiness. Traditional institutions in the solitary communities of the past had validated explanations as it had practices. Folk wisdom and the learning of the authoritative custodians of faith embraced all the accumulated knowledge of the group in a continuum that touched every aspect of experience.

Knowing was functionally related to doing. Men wished to know because they wished to be certain when they acted. The particularities of information or explanation were not critical; nor was the predictive value of any datum or concept. Minds that were not open to experiment did not seek to test the effectiveness of ideas – indeed, the very concept of doing so with reference to any of the great subjects of human concern was unthinkable. How could one verify the worth of the open-field system – or the efficacy of prayer? In these matters, men wished to know not in order to decide as between alternative courses of action, but in order to feel secure in the acts which tradition in any case dictated.

The community was aware that there were ways of knowing outside the tradition, free of its own oversight. It regarded them with suspicion as black mysteries, for they led to deviant actions and reflected the working of strange, perhaps unholy powers. The heretics, the infidels, the Gypsies, the Jews, the witches, and the magicians shared common attributes to the extent that their abnormal behavior was connected with illicit, that is, unsanctioned, knowledge. Therefore they were at once feared and hated.

Modern scientific enterprise was from the start suspect on somewhat the same grounds. Although it was long sheltered in established clerical institutions – the universities and the church – its practitioners were detached from any community but their own and their basic goals differed from those of other men. Whatever formal obeisance the scientists made to tradition, their ultimate quest was for change rather than certainty. They unsettled rather than confirmed accepted beliefs because truth was for them an end in itself rather than a means of explaining and justifying existing habitual practices. They possessed and exercised a magic, the character and purposes of which were unknown to the uninitiate. Science smacked of the forbidden; it threatened to uncover secrets that were best left concealed. Hence the widespread dread of what might come of it in the eighteenth and nineteenth centuries when it had acquired considerable independence and had developed institutions of its own.

Even though the underlying tensions about its effects persisted, as the nineteenth century progressed, unorthodox opinions in time encountered less opposition; and even the war with theology lost its point and its bitterness. The growing tolerance of science was a factor of its utility to new, but increasingly important, social groups. Entrepreneurial types, whether among the gentry or the middle classes, who valued personal achievement above ascriptive status and who regarded inherited tradition and the entrenched community as the sources of restraints upon their desire to fulfill themselves, were likely to be experimental and calculating in their attitudes, to esteem reason above habit as a guide to action. There was likely to be a high incidence of such men in the mobile populations of the growing cities, among marginal ethnic groups and within various dissenting sects; but personalities receptive to new ideas were by no means confined within those social limits. Increasingly, such persons found attractive the kinds of knowledge that science generated.

The expanding scientific enterprise appealed to its own practitioners on abstract grounds, as a mode of progressively uncovering the truth – a good in itself. But it also drew the support of a widening circle by its utility in improving the ways of doing, demonstrable in practical results. The claims of astronomy, for instance, were long since familiar in advancing the methods of navigation. Geology and biology, by the same token, were instruments by which to develop better techniques of mining and agriculture. Discoveries in physics and chemistry would have a comparable effect upon industry. The justification by utility became a conventional tactic of the nineteenth century.

The linkage of science with practice was clearest, most dramatic, and most effective in medicine. This was a field in which ways of knowing were intimately connected with ways of doing and one in which the welfare of every man was concerned. From the mid-nineteenth century onward, the conviction grew that the way to health passed through the medical laboratory. Since in this area the scientists had access to clinical materials only through practice, he had a stake in nurturing the belief, which brought him patients, that his ministrations would yield measurable results in the cure of illness. By the end of the century the association of hospitals, universities, and laboratories in a firmly articulated complex was the visible manifestation of the union of science and technology.

By then also comparable claims were being heard from newer disciplines, including the social sciences. Political and economic systems and family and interpersonal relationships of other sorts were subject to pathologies of their own, which the proper organization of knowledge could ameliorate. Science in this broadest sense took every human concern as its province.

Common to all these assertions of the preeminence of science was the assumption that every deficiency in man's world was definable as a problem to which the correct ways of knowing would supply an appropriate solution. The staggering optimism of this article of faith endured into our own times and endowed science with the vital force to sway the opinions of the increasing numbers of its clients. Ultimately it promised that the organized use of intelligence, through its procedures, would perfect man.

The conquest of opinion was never complete. Traditional folk wisdom retained its hold in the more stable communities and among the less adaptable personalities. Familiar beliefs and practices persisted. But they became the deviations from a norm which had secured broad acceptance in the society as a whole and which by the twentieth century were buttressed by a formidable array of institutions.

Significantly, however, a subtle foreboding about the consequences disturbed even the individuals most susceptible to the claims of the beneficence of science. Again and again, a half-admitted fear creeps into the imaginative efforts to envision the future. An early nineteenth-century version of a myth gave popular form to the dread that continued to trouble even men committed to science.

Frankenstein, a dedicated young scientist who seeks knowledge to help mankind, discovers the secret of life through the study of electricity, galvanism, and chemistry, and applies his formula to create a machine-monster. The monster, however, quickly proves himself the superior. In the confrontation, the machine gives the orders: 'Slave, I before reasoned with you, but you have proved yourself unworthy of my condescension. Remember that I have power. . . . I can make you so wretched that the light of day will be hateful to you. You are my creator, but I am your master; – obey!' The monster becomes the oppressive master of man, although it was neither evil to begin with nor created out of deliberate malice.

Mary Shelley later recalled the circumstances under which the idea came to her, in 1816, while she listened to her husband and Lord Byron pass the long evenings in talk by the shores of Lake Leman. Often their conversation turned to science, and particularly to the mystery of electricity and to experiments in creating life through galvanism. And a vision suddenly came to her of the dreadful 'effect of any human endeavour to mock the stupendous mechanism of the Creator of the world'. The impiety inherent in the magic of which they spoke invited retribution.

Mary Shelley gave her novel a subtitle, *The Modern Prometheus*. It was a theme that her husband had also treated, and that had recurred for centuries and would continue to recur in Western imaginative writing. In another form that theme also appears in retellings of the Faust legend of a demonic personality unwilling to respect the unfathomable.

Even the most optimistic imaginations long continued to shudder with that primitive fear. 'This accursed science', exclaims H. G. Wells, 'is the very Devil. You tamper with it – and it offers you gifts. And directly you take them it knocks you to pieces in some unexpected way. Old passions and new weapons – now it upsets your religion, now it upsets your social ideas, now it whirls you off to desolation and misery!' The ability to work miracles leaves the world 'smashed and utterly destroyed'.

Within this perspective of disrupted communities and changing ways of doing and knowing, the developments of the past seventy-five years become more comprehensible. There was always a grudging acquiescence in the victory of science and technology. Now the inner changes in the new ways, despite the magnitude of the achievements to which they contributed, further alienated the mass of men, and heightened the suspicion of the power thus created.

From the middle of the nineteenth century, the structural changes that isolated science from modern life proceeded rapidly and radically. Knowledge became specialized, professionalized, and institutionalized. The three tendencies were interrelated and each had the effect of creating a closed body of skills and information. Specialization was the product of forces inside and outside science. The mere accumulation of data stored up in libraries made it increasingly difficult for an individual to master more than a limited sector of any field. It seemed to follow that the more limited the field the more readily could it be mastered; and that, in itself, further encouraged specialization. In addition, the emphasis upon classification as the first step in all inductive learning induced scientists to mark out, and concentrate in, a distinctive and circumscribed field of research. Finally, the growing rigor of the tests for validation required constantly improved techniques; and it was a rare individual who commanded more than one set. Science – at any rate, for most of its practitioners – became the province of the expert who excelled in the one subject he knew thoroughly.

The result was a high degree of compartmentalization, the fragmentation of knowledge into a multitude of different disciplines, each familiar only to its own initiates. A chemist was not much more able to discourse with an astronomer than with a sociologist, or an economist with an anthropologist than with a physicist. The overlapping of techniques, language, and subject matter kept some lines of communication open, but each field was really known only to those who specialized in it.

Specialization contributed to the great advances of modern science. But it also demanded such a high degree of competence that it, in fact, excluded the amateur and made the practice of science entirely professional. Learning now required a prescribed course of preparation; it imposed defined canons of judgement and validation; and it developed the *esprit de corps* of a coherent and united group.

Finally, specialization and professionalization tended to institutionalize science. Research became so expensive that no individual could buy his own telescope, computer or cyclotron; and the organization of scientific enterprise increasingly fell into forms established by government, business, universities, and foundations. These characteristics facilitated the great achievements of the past century. They were also responsible for the developing gulf that set scientists off from other men.

The accelerating pace and audacity of discovery magnified the effects of the distance between the learned expert and the rest of the population. In 1900, the graduate of a European gymnasium or lycée or of an American college could expect the knowledge acquired there to retain some currency through his whole lifetime. In 1950, anyone who wished to keep abreast had to resist desperately a mounting rate of obsolescence; although far more people were educated than formerly, their schooling now equipped them with information that could quickly lose all value. Furthermore science grew ever less inclined to replace old with new certitudes; it ceased to deal with deterministic laws and yielded instead tentative statements of probability. At the same time, it probed the most important aspects of human existence and did so with increasing confidence. Since Darwin's day it had been busily destroying the fixed universe of tradition; now it made clear that it offered no consolatory alternative of its own.

The popular response was complex; people learned to tolerate but not to assimilate science. They accepted its judgements as true, since they were now validated as authoritative. But they blocked them out as irrelevant by refusing to adapt to them the beliefs or behavior of daily life. The ability to answer correctly questions about the new astronomy or physics or psychology did not modify old views about heaven and hell or about absolute personal morality. The two kinds of knowledge co-existed in uncomfortable juxtaposition.

Technology, however, made it difficult thus to isolate science from life. As the nineteenth century drew to a close, the new knowledge invaded industry, changed the machine, and altered the nature of the factory; it then impinged directly upon the experience of the laborer who could no longer escape an awareness of its implications.

The most striking indication of the transformation was visual. In the factories built in 1900, the drive shafts and the pulleys were no longer visible. Power was transmitted through wires and tubes – often hidden – and the whole was covered up and shielded so that the machine gave the appearance of being self-contained and autonomous. The onlooker no longer saw a comprehensible apparatus; he saw an enclosed shape actuated by a hidden source of power from which the products flowed by an occult process.

Some of the changes in design were incidental to other purposes. The demands of safety, for instance, often produced the shields that concluded the mechanism of operation. Other modifications were aesthetic, although even those were related to the meaning the machine held for men. An unbroken sheet of black metal seemed more pleasing to the eye than a complex of belting and gears because it conformed to the idea of the machine as self-contained.

The application to the machine and to industry of electricity was an even more important break with past human experience. Men had been experimenting with various manifestations of electricity for two hundred years, but it remained a mysterious force, somehow confused with galvanic magnetism, somehow related to the secret of life, but not popularly understood, not as comprehensible as water and steam power had been. Even after a multitude of appliances had brought it into every home, few men could grasp how a current passing through a wire created light and sound or turned the wheels of great machines. Still fewer would comprehend the processes involved in the technological application of knowledge from electrochemistry or nuclear physics to the instruments with which they worked.

The gap between the machines and their users widened steadily. In the twentieth century it was no longer the tinkerer who was inventor. The innovations were less likely to be products of industrial experience than of science; and the people who operated them understood neither the machines they served nor the technical fund of knowledge that dominated industry.

But they could not fail to be aware that their own conditions of life and labor were changing at the same time and in response to forces generated by the new technology. The factory became a new and different kind of human environment toward the end of the nineteenth century. The numbers involved were much larger than in any previous era. The plant no longer counted its employees in the scores, but in the thousands; and that increased its impersonality and rigid discipline. The analogy to the army became closer and more frightening as individual identity diminished in importance. The hordes that passed through the gates each morning had to be accounted for and their time put to a precise, measured, profitable use. Before the end of the century, Frederick W. Taylor had already outlined the principles of industrial management; and the demands of efficiency were served with increasing severity as the decades passed. Technological innovation became not only an end in itself but also a means of establishing greater control over the labor force. The more enlightened enterprises recognized the importance of human relations; but they did so as a means of increasing their efficiency, and the devices they used had the further effect of manipulating the lives of their employees.

An altered environment increased the external pressures upon personality. In the second half of the nineteenth century the factories had become urban, either through the growth in the size of the towns in which they had been located or through the shift to metropolitan centers. All the difficulties of industrial experience were therefore compounded. The machine, the factory, and the city became identified as a single entity oppressive of man.

The optimists consoled themselves with the vision of abundance of goods produced by the new regime and they hoped that the machine would compensate for the deficiencies of their own society by resolving in the future the problems with which they could not deal in the present. *The Shape of Things to Come*, as H. G. Wells saw it, was dominated by industrial plenitude that was a product of invention. Self-consciously the engineers assumed that they could not only make the machines run but manage society as well; and the technocrats envisioned a mechanical order – efficient, antiseptic and capable of dealing with any contingency.

Yet overtones of uneasiness about the mysterious new ways of knowing also persisted. In 1911, Henry James, visiting a laboratory walled with cages of white mice, exclaimed at the magnificence of the 'divine power' exhibited in cancer experiments. But he also wondered about the personality of the little creatures imprisoned in the wooden cubicles. Other observers also speculated about who was making the decisions and toward what end.

Doubts therefore always offset the confidence about the results. Karel Čapek's *R. U. R.* in the 1920s created a sensation in its nightmare vision of a robot's universal which completely dominated humanity. The greatest of the popular artists, René Clair and Charlie Chaplin, in *À nous la liberté* and *Modern Times*, expressed an identical protest: the assembly line made man its slave, repressed his emotions, and crushed his individuality. He could escape to freedom only by revolt against the machine. In *Brave New World* and *1984*, Aldous Huxley and George Orwell stood utopia on its head. The necessity for mobilizing large groups along military lines, which provided Bellamy with his Industrial Army, to these writers established a terrifying engine of oppression.

Much in these protests had a familiar ring; to a considerable degree these artists repeated the criticisms of the machine already voiced in an earlier, simpler era. But the involvement of science in the technology of the more recent period diffused the concern to much wider circles which had theretofore equably accepted change in the ways of doing but which now could not readily evade the need for new ways of knowing. However routine his role, the technician at the keyboard – or, for that matter, the indiscriminate television-viewer – had to wonder sometime what occurred within the box, the dials of which he turned.

Since the explanation of the scientists was remote and incomprehensible, a large part of the population satisfied its need for knowing in its own way. Side by side with the formally defined science there appeared a popular science, vague, undisciplined, unordered and yet extremely influential. It touched upon the science of the scientists, but did not accept its limits. And it more adequately met the requirements of the people because it could more easily accommodate the traditional knowledge to which they clung.

Pragmatically, the popular science was not always less correct than the official. It would be hard to assert with confidence, for instance, that faith healing, nature cures and patent medicines were always less effective than the ministrations of the graduates of recognized medical schools; or that the vision of the universe exposed in the television serial or Sunday supplement was less accurate than that of the physics textbook. By the tests of practice – of whether it worked – popular science did as well as official science.

The deficiencies of popular science were of quite another order. It formed part of no canon that marked out its boundaries or established order among its various parts. It consisted rather of discontinuous observations, often the projection of fantasy and wish fulfillment, and generally lacking coherence and consistency. Above all, it embraced no test of validity save experience. It was as easy to believe that there was another world within the crust of the earth as that there were other worlds in outer space. One took the little pills; the pain went away. One heard the knocking; the spirits were there. The observable connections between cause and result were explanation enough. It was unnecessary to seek an understanding of the links in the chain between the two.

The men and women who moved into the highly complex and technically elaborate industrial society of our times simply assimilated the phenomena about them in terms of the one comprehensible category they already knew, that of magic. And it was thus too that they understood the defined science of the laboratory and the university. The man who pressed a button and saw the light appear, who turned a switch and set the machine in motion, felt no need to understand electricity or mechanics; the operations he performed made the limited kind of sense that other mysterious events of life did. The machine, which was a product of science, was also magic, understandable only in terms of what it did, not of how or why it worked. Hence the lack of comprehension or of control; hence also the mixture of dread and anticipation.

As in the past, the new ways of knowing seem unconnected with tradition, appear to be the possession of outsiders, and are considered potentially dangerous. But for people who fly in jets and watch television, the threat has an imminence it did not have earlier. Those who blankly and passively depend upon modern technology frequently feel themselves mastered by science without knowing why.

The more useful science becomes, therefore, the more it is both respected *and* feared. That it had power at its disposal was all along known; now the machine compounds its force and creates the suspicion that it is buying men into bondage. The people who are simultaneously delighted with additional years of life expectancy and terrified by the bomb are in no position to strike the balance of their gains and losses in happiness. In their confusion, they wonder whether the price of the gadgets which delight may not be servitude to the remote and alien few who control the mystery.

The popular response to science is thus ambivalent, mingling anger and enthusiasm, lavish support and profound mistrust. Conceivably the tension can persist unresolved indefinitely into the future as it did in the past, although it would be hazardous to count on that outcome under rapidly changing modern conditions.

The profound uneasiness about the consequences of the new ways of knowing will be quieted only if science is encompassed within institutions which legitimate its purposes and connect its practitioners with the populace. Education is helpful insofar as it goes beyond the diffusion of techniques to familiarize broad segments of the population with the basic concepts and processes of science. But even those who never acquire that understanding need assurance that there is a connection between the goals of science and their own welfare, and above all that the scientist is not a man altogether apart but one who shares some of their own values.

Reference

HANDLIN, O. (1964), 'Men and magic: first encounter with the machine', *American Scholar*, no. 23, p. 408.

15 S. B. Barnes[1]

On the Reception of Scientific Beliefs

Although this paper is concerned solely with scientific beliefs, its aim is neither to describe singular features involved in their reception, nor to develop a specific theoretical approach to this. Rather it seeks to show that the application of general theoretical principles, largely accepted in the social sciences, can lead to interesting conclusions in this area – and, additionally, can expose inadequacies in existing work stemming precisely from an exaggerated stress of the special status of scientific beliefs.

In conformity with this strategy it is appropriate to start by considering the explanation of beliefs in general. There is no complete consensus, as yet, in the social sciences, upon the best approach to this problem; indeed the meaning of 'explanation' in this context is not finally settled. It is, however, possible to codify explanatory strategies widely followed in practice, and to demonstrate their relationship to each other. This will be attempted here in a formal and abstract way, starting from an action frame of reference. An interest in beliefs comes close to entailing this starting point.

When we attribute beliefs to actors we assume a general consistency in their linguistic usage and in the relationship between that usage and their actions; without this being the case we could not establish that actors had a language or held beliefs. (See Lukes, 1967.) Given this it is possible to gain an understanding of the meaning of actors' beliefs by 'empathy'. Only in this way is it possible to relate beliefs to actions, and to other beliefs; the significance of a particular belief is understood as the total system in which it resides is understood. And understanding is necessary

1. A slightly different version of this paper was presented to the Edinburgh University Social Sciences Seminar on Science and Technology, in May 1970.

before explanation can start. If beliefs are held to be of theoretical interest, then there is little alternative to this 'action' approach; it implies that an actor's concepts and beliefs must be identified and understood within his own framework, his own explanations of them appraised, before they become the data of any external explanatory theory.

Explaining beliefs

It remains to be shown how a sociological explanation of beliefs can be developed from this approach.[2] Let us start with an individual actor, loosely characterized by preliminary empathetic study, as holding a belief about a matter of fact. How is the belief to be accounted for? The path to an answer is most clearly presented as a series of formally differentiated stages, the first of which must be investigations of:

1. The extent and nature of the information potentially available to the actor.

2. The criteria by which the actor decides what information is relevant to his belief and how that information is to be interpreted.

On completing the above[3] we can usually class the belief, in the light of information available to the actor, as compatible or incompatible with his own criteria; we may describe the belief as consistent or not. Clearly belief in a flat earth, where the actor employs criteria of relevance and interpretation of information similar to our own, can be consistent if there is limited available information. Again, if an actor believes that grass snakes are poisonous, and we discover that he read this in a book whose statements he regards as good indicators of truth, then the belief is consistent. (Equally we might have related a child's belief in atoms to statements in the 'Boys' Book of Wonders'.) Consistent beliefs can be explained economically as part of the actor's normal practice; thus the preceding example can be explained, along

2. Although I shall frame most of the arguments in this paper in terms of beliefs it is with the proviso that beliefs and actions are only meaningful in terms of each other and cannot be independently investigated.

3. This formulation avoids discussion of a number of methodological controversies, notably those centering on the relationship of the beliefs actors profess to have and the beliefs we may infer that they have.

with a whole class of beliefs transmitted by the book referred to, in terms of a single rule – the existence of which may itself of course demand explanation. However, inconsistent beliefs, which apparently deviate from the actor's own rules, cannot be explained in terms of them. The next stages in the explanatory sequence are relevant only when beliefs are inconsistent yet the actor cannot be brought to acknowledge this. Two approaches are possible here:

3. One possibility is to assume that the actor's rules and concepts have been incorrectly understood. When the actor's language is either different from our own, or involves esoteric vocabulary, we can explore whether or not an error of translation has occurred. When the language is apparently the same as our own we can investigate differences in usage. In this case an identity in vocabulary disguises real differences in meaning. A belief about solid bodies, for example, may seem inconsistent only because the actor's use of the term 'solid' is different from our own.

A complete investigation of the meaning of a term through usage involves examining a wide variety of situations; the significance of these situations can only be understood in terms of other meanings; inexorably one is required to obtain a full empathetic understanding of the whole way of life of an actor in order to exhaust the possibilities of this approach: in practice the possibilities are never exhausted.

4. The other strategy we can adopt with regard to inconsistent belief is that of causal explanation. The complex of causal influences surrounding an actor normally favours consistency with the actor's own rules; in some circumstances however special causes of inconsistency may be present. Social and cultural factors are often cited here; the demands of a religious faith may result in the inconsistent denial of a proposition; the protection of an interest may account for the retention of an inconsistent belief. In any situation an enormous number of factors which might predispose to inconsistency will invariably be present; their influence will rarely be acknowledged by actors; thus the possibilities of this second strategy are as difficult to exhaust as those of the first.

Inconsistent beliefs then are explained by reduction to consistency or relation to causes. In theory concrete applications of these alternatives can generate conflicting testable hypotheses; in practice however, as indicated above, it is not possible to push either approach to final refutation. The consequence of this has not been a tentative eclecticism, instead investigators have tended to make a commitment to one or other approach. (The first has facilitated justification, the second criticism of the beliefs in question.)

It remains to explain the rules and standards of consistent action. These are themselves beliefs and their status as standards of consistency is conventional rather than necessary; in other contexts their consistency may be judged by reference to other beliefs; consistency springs from the systematic interconnectedness of beliefs and actions. None the less we may still approach these rules and standards deterministically; there is no reason why we should not demand a causal account of how an individual actor came to adopt them.[4] To meet this demand is the task of the next steps in the explanatory process; the first step is an acknowledgement of a generalization firmly established by empirical sociological and anthropological work:

5. The human actor adopts a way of life largely determined by his culture and the position he occupies within it; many of his beliefs and most of those crucial to the acquisition of further beliefs, will be found empirically to have been received in socialization processes. Theories of the socialization process will eventually provide the answers to most problems in the adoption of beliefs.

In some cases however a belief may be adopted despite a contrary cultural indication, in the absence of any such indication, or where there is a real choice (even a necessary choice) of available beliefs. Two forms of explanation are commonly used in such cases:

6. We can attempt to demonstrate how the actor, *with a given system of concepts and beliefs*, is influenced by his experience of the world into adopting new beliefs or generating new concepts.

7. Alternatively causal analysis can be attempted similar to that described in 4. Social and cultural factors may act in special ways to influence the adoption of specific beliefs.

4. There is a school of thought which would exclude causal analysis entirely from the social sciences. See Winch (1958).

As with 3 and 4 these two forms of explanation may compete or cooperate in the explanation of whole areas of belief. Of course once a new belief is initially adopted its further proliferation becomes increasingly influenced by group pressures and mechanisms of passive reception; we are led back to the socialization process as an element in explanation. But as most new beliefs compete with and displace old ones, and the old ones will normally dominate existing socializing contexts, causal and experiential factors can be held to play a major role in the spread of beliefs through societies, as well as in initial individual acts of adoption; thus we can discuss social conditions appropriate for the rise of a particular pattern of beliefs.

It has been implied that individual actors receive most of their beliefs through socialization and are actively involved in the choice, modification, rejection or creation of but few of them. This selective and creative activity, however, performed by many actors over a long period, can result in substantial changes in the beliefs of a culture as a whole. This leads to the final stage in the explanatory sequence:

8. We must explain why, in the long term, some beliefs persist as stable features of a culture, as institutions, whereas others are progressively modified or eliminated. These emergent problems form part of the range of questions with which the classic theories of sociology, such as functionalism and Marxism, have attempted to deal.

Starting from curiosity about a particular belief held by a particular actor I have attempted to reveal the complexity of ensuing explanation and the way it requires consideration of the full social context. Simplification of the scheme is sometimes possible; some stages of investigation may be left implicit when actor, social scientist and reader share concepts and beliefs to a very great degree and enjoy identical perceptions of the relevant situation. However, awareness of the scheme (or rather of the problems it articulates, for the actual presentation here is grossly inadequate and oversimplified) is particularly important when the 'actor's point of view' is likely to be considerably at variance with that of the investigator. In this regard social anthropology and the sociology of science share a common difficulty, and in both fields failure to appreciate actor's beliefs in their own contexts, and a tendency instead to evaluate them by external criteria, has had averse effects. In anthropology the difficulty has come from treating beliefs differently according to their 'truth' or 'falsity' in external terms; in the sociology of science similar difficulty has come from treating the 'truth' of scientific beliefs as theoretically interesting.

Simply to label beliefs as objectively 'true' or 'false', in itself does not cripple a theoretical approach to their explanation; this happens because 'true' beliefs are sometimes held, in the social sciences, not to require explanation at all. Curiosity is diverted from beliefs consonant with those venerated in our own society (and hence clearly 'true'); it focuses only upon 'erroneous' beliefs, and then the question is not so much why they are believed as *whyever* they are believed. This is clearly incompatible with an 'action' approach which explains beliefs through their significance for the actors holding them, not for the criteria of some unknown external system. Peel (1969) has discussed the way in which anthropologists have confined causal analysis to 'erroneous' beliefs: he writes –

> . . . it may be a true biographical fact about people in general, that they only do want causal or genetic explanations of beliefs which they consider false, and regard their own beliefs as somehow supra-historical . . . but this is not a logical connection. True beliefs as well as false ones are the product of social forces and their origin is a perfectly legitimate concern for the sociologist; causal explanation is not to be restricted to what the sociologist's own society considers false. Indeed in view of the fact that, for example, mistaken cosmologies have at most times and places been more prevalent than our present 'scientific' cosmology, the origin of the latter, in time and place, demands explanation. . . . It is not legitimate for the scientist or rationalist to exempt his views from sociological analysis by saying 'I hold these views simply because the evidence demands it'; we want to know why he, of all people, has come to interpret the evidence in this way. Conversely, to give a causal account of a belief is not to undermine its validity; it just is not relevant to it. (p. 71.)

Peel is able to show how this failure to fully relativize – to treat our own beliefs as objects of study like those of others – results in inadequate explanation of the beliefs of other peoples. A similar failure is found in sociological analyses of scientific beliefs and their reception; there is a common presupposition here that scientific activity leads spontaneously in a straight line to truth, unless external causes block or divert it.

Science, error and irrationality

Barber (1961) has considered the way in which a number of scientific theories now accepted as correct were initially resisted by scientists; Mendel's theory of inheritence, Pasteur and Lister's theories of germs and Planck's views on quantization are cases studied. His technique is to search for social and cultural biases that may be held to explain the 'errors' involved; religious commitments, the maintenance of disciplinary boundaries, the effects of status and prestige, theoretical, methodological and conceptual commitments and loyalty to a particular 'school of thought' are biases cited. After listing instances of resistance, each accompanied by plausible causes of error, Barber concludes:

> That some resistance occurs (within science), that it has specifiable sources in culture and social interaction, that it may be in some measure inevitable, is not proof either that there is more resistance than acceptance in science or that scientists are no more open minded than other men. On the contrary, the powerful norm of open-mindedness in science, the objective tests by which concepts and theories often can be validated, and the social mechanisms for ensuring competition among ideas new and old – all these make up a social system in which objectivity is greater than it is in other social areas, resistance less. ... Nevertheless, some resistance remains, and it is this we seek to understand and thus perhaps to reduce.

Clearly the investigation is radically different from the ideal delineated above. In no case is evidence for beliefs considered, or its availability to actors; where actors' concepts and evaluative beliefs are mentioned it is to present them directly as erroneous, or accept them directly as objective, in the light of current scientific orthodoxy; explanation consists entirely in finding causes for externally defined error.

The difficulties of this approach are readily made apparent. The acceptance of theories we still regard as correct has been favoured by social factors: neo-Platonism, for example, provided a climate of opinion important in the favourable reception of the Copernican cosmology. Conversely many views objectionable in terms of current ideas have been kept at bay by 'preconceptions' or the low status of their proponents. (See for example, Polanyi, 1957 and 1958.)

Just as we may doubt whether 'truth' is always the result of 'objectivity', so may we wonder whether 'error' always stems from 'bias'. At the time when evolutionary biologists were battling with the 'irrationality' of colleagues within their discipline, their views were criticized by Lord Kelvin, the physicist; he calculated that the sun had contained insufficient energy to heat the earth for a period long enough to allow the evolutionary processes to take place; the argument was impeccable in terms of the physics of the day; 'truth' triumphed by ignoring it for thirty years. This leads to a further criticism: since the truth defined by science is perpetually changing, a theory such as Barber's would need to make embarrassing reappraisals of the causes it cited, degrading the importance of one scientist's religion to stress instead the authoritarian upbringing of another, and so on.

For Barber true and false conclusions arise respectively from evaluations in conformity with scientific method and those made in defiance of it. Theories can always be evaluated immediately because the scientific method is thought of as simple, readily applied, highly general and sufficient in itself to establish validity. These assumptions come readily to Barber, whose work belongs to a tradition which regards the lengthy and esoteric training of scientists as consisting in methods of obtaining truth; this, once obtained, is readily identified. From this viewpoint the scientist is like a miner who applies sophisticated skills to the extraction of gold, which, once obtained, can be immediately recognized by simple tests; simple tests for truth used by the scientist involve open-minded appraisal of its 'consonance with experience'. This analogy is a rewarding one, for the truths of science are, for Barber, like gold, permanent and immutable; throughout his work he studies beliefs comprising underlying highly general, explanations for phenomena; yet these are referred to not as theories but always as *discoveries*.

In the terms of reference previously defined, Barber regards the application of scientific method as the universal 'normal practice' of science, completely specifying the nature of its evaluation processes; deviation from it is explained in causal terms. Present scientific truth, he assumes, would always have been arrived at, if scientific method had been applied in the past, hence it acts as a reliable indicator of whether the method was applied or not. In this, both the role and the significance of scientific method have been misrepresented.

The presuppositional nature of science

Belief in the scientific method and the attendant assumptions described above, which fit so naturally in research programmes limited to explaining error, form part of a viewpoint which has for a long period been dominant in sociological, historical and philosophical approaches to science. Recent work has however moved away from this view. Particularly important in this trend have been writers concerned to describe concretely the actual process of scientific investigation and evaluation: the theoretical positions of Kuhn (1962, for example) and Polanyi (1957 and 1958) are both embedded in such concrete material, and both conclude that far more is involved in scientific evaluation than application of general criteria derived from scientific method.

Polanyi stresses the way in which esoteric beliefs and standards are tacitly involved in all scientific judgements and points out that in many instances scientists are not able to render these tacit elements explicit. The hypothesis that mammalian gestation periods are all integral multiples of π days would, Polanyi points out, never be taken seriously as a scientific generalization; it is, apparently, 'consonant with experience' but this is of no significance when it clashes strongly with tacit elements of biological judgement. Conversely, favoured beliefs, such as were involved in the theory of the partial dissociation of electrolytes in solution, may not be discarded even when confirming evidence is outweighed by anomaly.

Kuhn, too, stresses the presuppositions inherent in scientific evaluations. Concerning scientific education he points out how readily processes involving dogmatic transmission of concepts, ways of looking at things, esoteric problem solving techniques and so on, can be identified. These elements together form a disciplinary paradigm which mediates the scientist's perception of his experience. What is to count as a problem, a solution, a successful experiment, a failure, a discovery or an anomaly, all are mediated by the received paradigm which involves an entire system of presuppositions to which the scientist is committed.

This work can in no way be regarded as cataloguing massive deviation among scientists; both writers are concerned to demonstrate the *essential* role of presupposition in science. Polanyi and Kuhn illustrate a general trend here, manifest in philosophy as an attack on empiricism and the idea of a neutral observation language, in history as a revolution in method, and in sociology by recent awareness of the importance of technical norms in evaluation processes. (See for example, Hanson, 1958; Hesse, 1965; Toulmin, 1961; Kuhn, 1957; Agassi, 1963; Mulkay, 1969.)

Presuppositionalist views of science command assent through the way they can come to grips directly with concrete scientific practice, but such assent involves abandoning scientific method as a sufficient source for external evaluation of truth and error. Once this notion of sufficiency is rejected the evaluation of scientific beliefs has to be seen in a new light. If there is no royal road to scientific truth, but instead a way only to be found by applying complex, technical standards of evaluation, then bias and irrationality are not the only reasons why one might get lost. And if these technical standards are recognized as contingent foci of commitment, involving no necessity, then it is arbitrary and misleading to treat an actor who does not hold to them as being lost. Patterns of meaning and standards of evaluation can no longer be taken for granted in studying scientific controversy, whether intra or extra-institutional; full examination of actors' points of view becomes necessary. As we are interested only in how actors evaluate the claims of scientists it is as well, in such an investigation, temporarily to suspend our own faith in their truth.

Overall, the implication of this analysis is that the sociological study of controversy and 'resistance' in science is particularly difficult. To empathize with the actors involved, and understand what counts as normal rule following, involves appreciation of an outlook normally imparted only through prolonged programmes of formal socialization and a familiarity with particular research situations difficult to acquire from the outside. Moreover, as different scientists follow different rules of action and utilize different esoteric concepts, the degree of their mutual comprehension cannot be assumed to be high.

When information is transmitted between scientists its background of justifying argument can be at any level of intelligibility. Low background intelligibility is characteristic of much interdisciplinary communication: many chemists and biologists using spectroscopic techniques accept the results of physical theorizing unintelligible to them; many workers in the life and earth sciences are similarly required to take on trust esoteric techniques for determining chemical structure. The consensus of one discipline may form legitimate knowledge for another. Even within disciplines intelligibility may be low, due to segmentation or other types of division of labour. And as no scientist can operate entirely upon the basis of information produced within his own subspecialty the trust of authoritative sources is a matter of routine.

When intelligibility is high, say within an 'invisible college', the picture is rather different, although descriptions based on the importance of complete 'openness' are still not appropriate. Of the work he utilizes, no scientist personally checks more than a small fraction, even of that he is fully competent to evaluate, nor, in general, is any responsibility laid upon him to do so. Instead more limited norms of scepticism exist: critical attention is directed to work anomalous with respect to a paradigm, results produced by a recently developed technique, and experiments either charged with technical inadequacy or unavoidably unreliable due to external constraints or dependence on measurements near a technical theshold. Results produced by practitioners reputed to be technically unreliable are given a similarly guarded reception whereas other work is likely to be accepted routinely, especially if it emanates from a highly reputable source. On logistic grounds alone such channelling probably makes good sense; without it work would proceed very much more slowly, and the consequent gain in reliability would be minimal.

Every scientist operates on the basis of an extensive matrix of existing beliefs. This matrix includes both his received paradigm and his evaluations of knowledge sources both written and personal. Beliefs within the matrix are held with varying degrees of tenacity and self-consciousness; experience mediated by the matrix as a whole may cause the scientist to alter some part of it; his own work may serve to admit him into the matrix of others as a trusted knowledge source within a defined area. The scientist does alter his beliefs in the light of experience but his normal response is not automatic and determined as that of a counter to radiation. (Nor is the analogy with a computer responding to an input in terms of its programme altogether adequate; this latter analogy is appropriate only if one allows that one possible outcome is a change in the programme itself.)

There is nothing about this description exclusive to the scientist; it could be argued that it fails to stress the most interesting and distinctive features of scientists' behavior, the ones most in evidence when they are being particularly original and hence presumably least influenced by their existing matrix of belief. Arguments along the lines of the above prejudge an issue in need of investigation. There is evidence that originality can spring from commitment to received beliefs, despite opposition from a changed reference group and the demands of new subject matter. (See for example, Ben-David, 1960.) On the other hand there is, as yet, no evidence of a connection between scientific originality and 'open-mindedness'.[5]

Barber's attribution of causes to the beliefs of scientific 'resisters' implied scientific inadequacy. Little work of this sort exists, possibly because of a reluctance to challenge the dominant image of the infallibility of scientists. More work exists attributing causes to lay resistance to the beliefs of scientists, and the same implicit judgements are to be found in this, unsupported by any insight into the actors' points of view. Here too examination of problems of intelligibility and patterns of trust can be useful.

Lay responses to scientific beliefs

Sociological studies of the lay response to scientific beliefs are often influenced by a preconceived vision of science's gradual triumph over prejudice and error; distortion of lay beliefs is a common consequence of this. It is interesting to note how often these beliefs are simply condemned as 'irrational'; the word seems, in practice, to be used as a stronger condemnation than 'erroneous' to

5. This is not to suggest that the way in which beliefs are held is totally unrelated to scientific effectiveness. Clearly one who never modified any element in his matrix of belief would be unlikely to succeed in science.

stigmatize deviations from particularly sacred beliefs, including those of natural science. Expansion of such criticism is rarely to be found; its meaning is taken to be obvious.

One clear example of this is to be found in studies of the fluoridation controversy. Fluoridation of the water supply has been recommended on grounds of dental health and nearly all the scientists involved are agreed that the process is completely harmless. Opponents of the measure claim, among other things, that the fluoride would act as a poison, and therefore that the measure should not be implemented. Much study of this issue has been directed towards accounting for the 'irrationality' of the anti-fluoridationists. In a study giving admirable attention to the beliefs actually held by anti-fluoridationists, Mausner and Mausner (1955) talk of the 'bizarre irrational fears' inspired by the belief that fluoride is poisonous, and later, urging more study of the beliefs and motivations of 'anti-intellectual' movements, they conclude: 'If the fluoridation case is any indication we shall find the motives understandable, if irrational'.

Irrationality, even more than error, conventionally demands causal explanation; Mausner and Mausner suggest emotional and unconscious sources for the beliefs, others suggest that feelings of alienation are responsible. Although these causal explanations are possibly valid, the need to label the relevant beliefs as irrational before applying them certainly distorts the situation. Irrationality implies, if anything, inconsistency in rule following, or indifference to experience and argument; nowhere is this demonstrated; there is not even an attempt to show that the anti-fluoridationists deviate from any ideally conceived scientific method in the way they appraise the claims of scientists; instead scientific authority is quoted and the actors are shown to be 'mistaken' with respect to it. Basically the 'irrationality' of the anti-fluoridationist is no more than the acceptance of a different authority source to the investigator and (presumably) the majority of his society; most trust the scientific consensus, some don't; deviance has perhaps been demonstrated – but irrationality?[6]

6. 'Irrationality' in the sense of departure from normal practice might have been plausibly demonstrated by further research. Thus some who disbelieved the scientists in this case may usually have regarded science as authoritative.

Similarly, Parsons frequently separates the rational, irrational and non-rational without elaboration of the way in which the feat is performed:

> ... there are many elements of ambivalence in public attitudes towards science and the scientist, which are expressed in much irrational and some relatively rational opposition to his role.
>
> The obverse of this is that there is a strong non-rational element in the popular support of the scientist. He is the modern magician, the 'miracle man' who can do incredible things. (1951, p. 339.)[7]

(In the context of his general sociological theories Parsons does, of course, discuss the concept of rationality. But, in practice, his criteria would produce different classifications of action in the hands of different investigators. In concrete analyses Parsons rarely attempts to show how he discriminates the 'rational' and the 'non-rational'; this is, I suggest, because in these situations he discriminates subjectively.)

We find greater awareness of the difficulties involved in the concept of rationality in Rogers' work (1962) on the diffusion of innovations.[8] Rogers discriminates both those who adopt and those who reject innovations into 'rational' and 'irrational' groups; he discusses this procedure as follows:

> It is often difficult to determine whether or not an individual *should* or *should not* adopt the innovation. The criterion of rationality is not easily measurable. The classification can sometimes be made by an expert on the innovation under study. Often, the classification is made on the basis of economic criteria.

Rogers favours objective definitions of rationality quoting with approval Homan's well known formulation (1961): 'Behavior is irrational if an outside observer thinks that its result is not good for a man in the long run.'

7. The term 'rational', whether neat or negated, hyphenated or otherwise amended, continually punctuates most of Parsons' work, usually without elaboration. The views expressed in this essay are clearly incompatible with the whole of Parsons' sociological framework.

8. It is interesting to note that corresponding to faith in the 'obvious truth' of natural science is a faith in the 'obvious goodness' of technological innovations. Writers in this field are generally aware of the relativity of 'goodness' in this context and acknowledge it in theoretical discussion, but subsequent typologies of 'causes of resistance' have clearly given little heed to it. See for example, Bright (1964).

But is there really value in a definition that may vary with the outside observer chosen? There is convenience in this, of course, if the outside observer is oneself. The force of Peel's previous argument (1969) should now be apparent as should the relevance of his conclusion concerning use of the term rationality:

> Judged from the standpoint of sociology no behavior is, properly speaking, irrational, for he who speaks of 'irrational' behavior speaks not as a sociologist trying for once to hold himself neutral with respect to different evaluative and cognitive standpoints, but as a partisan of a particular social viewpoint, or as 'a publicist, a polemical writer of tracts for the times'.

In practice at least 'rationality' acts as an evaluative not an explanatory term; a genuine search for explanation must necessarily return to taking actors' points of view; where a belief is in dispute, the concepts and standards of all actors involved must be examined even where some of them are scientists. Scientists' beliefs will on the whole derive from the esoteric concepts, taxonomies, images and working rules embedded in their disciplinary paradigm. These can be understood, as the scientist understands them, only in terms of each other and the practice in which they are embodied; a minimum of familiarity with the appropriate discipline is needed, considerably exceeding that possessed by most of the laity. The lay actor will accordingly find scientific theories and their justification for the most part unintelligible.

One strategy that can be adopted in these circumstances is for the actor to make the material intelligible as far as possible in terms of his own concepts; it will be rare indeed that no shift of meanings occurs in the process. Mostly this can be regarded as misunderstanding on the part of the lay actor; in some circumstances it is more profitable to view the same shifts as metaphorical extensions of the original material. More commonly, however, actors recognize that it is impossible for them to evaluate scientific discourse and today they generally take scientific conclusions on trust. In many cases the reception of scientific beliefs can be treated simply as the response of actors to an institutionalized knowledge source.

Scientific disciplines as legitimate sources of knowledge

Whereas belief in simple, highly general tests for truth leads to stress being laid on the role of argument and 'rationality' in the reception of scientific beliefs, awareness of the esoteric presuppositions essential to scientific evaluation illuminates the value of investigating scientific disciplines simply as accepted knowledge sources, and asking what factors contribute to or detract from their effectiveness in this regard. I shall argue that the success with which scientific disciplines can operate as general social sources of knowledge is limited by three of their characteristics:

Knowledge sources are most effective when they present an image of unanimity and certainty; actors, lacking other guides, use these as indicators of reliability. Scientific disciplines, however, lack institutionalized mechanisms for mobilizing and maintaining a consensus to present to an external audience.

Actors oriented to a scientific discipline as a knowledge source cannot perceive its internal boundaries of competence and allocations of status. This can strengthen the external influence of views discredited within science.

The strong internal consensus achieved by scientific disciplines and essential to them, is maintained by a stringent narrowing and purification of agreed goals within them. Consequently both institutional and individual constraints have developed discouraging the involvement of an entire discipline with specific social problems.

The effect of these factors is best demonstrated by example: particularly valuable work to cite here is that by Reiser (1966) on the United States Congressional response to scientific advice on a public health issue of considerable importance. Throughout the 1950s a number of papers and reports appeared claiming that a strong relationship existed between smoking and lung cancer. In 1962 the claims had achieved such prominence that Federal and private interests in the USA agreed on the appointment of ten highly regarded and hitherto non-involved scientists to report on the strength of available evidence. This advisory committee reached a unanimous conclusion (Public Health Service Publication no. 1103):

Cigarette smoking is causally related to lung cancer; the magnitude of the effect of cigarette smoking far outweighs all other factors. In comparison with non-smokers, average male smokers of cigarettes have approximately a nine to ten-fold risk of developing lung cancer and heavy smokers at least a twenty-fold risk. (p. 31.)
. . . cigarette smoking is a health hazard of sufficient importance in the United States to warrant appropriate remedial action (p. 33).

The result of this unanimous and uncompromising statement was a series of Congressional hearings in 1964, preliminary to intended legislation. In the event this legislation proved extremely feeble and reflected, according to Reiser, doubt of the validity of the scientific evidence. How did this come about? One important point was that a small minority of testifying scientists flatly rejected the findings of the advisory committee. Another was that the group supporting the findings exposed a number of technical differences of opinion. Although they had, it seems, realized the importance of unanimity and decisiveness on the main point at issue, they had not anticipated the possible adverse consequences of disagreement in other areas; the processes of disputation and criticism normal within their fields did not increase their credibility in this context.

The advisory report had cited both experimental and statistical studies of smoking; the latter however comprised most of the evidence. The scientists held that strong correlations between smoking and cancer incidence were strong evidence for a causal connection from the former to the latter; theoretically however they admitted that the correlation might result from a third factor causing both smoking and cancer. While they regarded the alternative theory as implausible and out of accord with other evidence, the scientists agreed that its conclusive refutation would require long term experimentation on controlled human groups. The notion of scientific belief being founded on massive probability seems to have been a surprise to the legislators, whose faith in the experts was further weakened when controversy concerning the nature of causality in science attained prominence at the hearings. Another cause of doubt was the way in which experts could declare their certainty that smoking caused cancer, and yet simultaneously agree that further research work needed to be done. The legislators had no perception of the details of the causal mechanism as an interesting scientific problem; that *their* problem still seemed to be of scientific interest surely indicated that it had not yet been solved.

The result was a strong recommendation that more research be carried out and tentative recommendations for feeble legislation. This was the outcome of many hours of detailed discourse on a (scientifically) fairly straightforward problem. In terms of the scientific orthodoxy the legislators erred in their appraisal of evidence; but to talk of error is pointless here; the legislators were confronted with a divided knowledge source whose standards they could not appreciate and whose conflict discouraged trust. Causal factors may have influenced the level of trust – legislators' smoking rates or the tobacco interest in their electorates – but who is to say what was the correct level?[9] And if views of the 'rational' triumph of scientific ideas do not apply to such a thoroughly argued case as this where can we expect them to apply?

To destroy the unanimity backing scientific evidence is a deliberate technique employed by some pressure groups. The anti-fluoridationist movement is a case in point; whether or not its activists are anti-scientific, they certainly give great prominence to any 'scientific' support they receive. The names of qualified scientists supporting their objectives (even if solely on ethical grounds such as individual freedom) are distributed with appropriate degrees prominently listed, and any statement by a scientist supporting the view that fluoride is harmful tends to be published with full attribution. Harvey Sapolsky (1968) has related the success of the anti-fluoridation movement to this technique:

> We argue then, that the voters reject fluoridation not because they are alienated but because they are confused. The public, believing the fluoridation controversy to be a conflict among scientific experts, seeks the safest course. Unable to decide between what appear to be two contending scientific positions, the voters opt to avoid the greatest potential health risks by defeating the fluoridation proposals.

Sapolsky supports his claim by citing a survey where 83 per cent of voters stated that they would vote for fluoridation if 'fluoridation was shown to be perfectly safe and able to help people's

9. Causal factors may also have affected the level of *professed* trust and made it markedly different from the real beliefs of the legislators. Unfortunately Reiser's methodology cannot cover this point.

teeth'. This must have come as a surprise to most interested scientists, who viewed the issue as settled and scientifically watertight. But, as pointed out above, 'insiders' and 'outsiders' operate with different beliefs and different amounts of information; Sapolsky's voters, looking to science for guidance, were unable to perceive its internal boundaries of competence and informal allocation of status. A list of those of their peers who maintained that fluoride had been shown to be harmful would be unlikely to impress most scientists; they would note an unusually large proportion of signatories from 'distant' specialties, scant representation of those, in any field, with significant and recognized research contributions, and an unusual proportion of those who had acquired notoriety elsewhere. The lay actor, however, confronted with tabulations of B.Sc.s and Ph.D.s running into three figures, is likely to be influenced, if he trusts scientists at all.

Careful of their external image, scientific communities rarely undertake public exposés of those they regard as incompetent; informal communication usually ensures that their work is treated as suspect, or in some cases written off. This method of control can cause embarrassment when a deviant individual occasionally explodes into the public eye. Statements from someone apparently accepted within a scientific discipline for many years gain weight accordingly, and, if concerned with the dangers of (say) nuclear fallout, fluoride, pollution or forms of accepted medical practice, they can cause considerable concern. No institutionalized processes exist in academic science to deal with this type of externally oriented deviance; scientists tend to sanction it in an ad hoc manner, occasionally bringing odium upon themselves with hastily conceived responses.[10] Since internal relationships are of overwhelming importance to the academic scientist, and an anti-dogmatic ideology is requisite to their maintenance, no scientific equivalent of 'unfrocking' has developed.

It is significant that the above examples both involve medical

10. See for example de Grazia (1966) and Sapolsky (1968). Sapolsky describes how a scientist claimed at a public meeting that dissenting colleagues had no 'right' to speak on the matter of fluoridation. It seems clear that 'right' in this context was intended to mean 'competence'. Unfortunately its significance was misinterpreted, and presumably served to reinforce the views of those suspicious of scientific 'control'.

research; this is the only large area where the process of extending knowledge is strongly subordinate to a specific ethic which embodies social values and is, on the whole, internalized by the scientists involved.[11] Medical research occurs in a field where a tolerable general value consensus can still be achieved, even in the highly differentiated societies of today. Most scientific disciplines investigate areas suffused with a multiplicity of objectives and evaluative criteria; as mentioned above, internal consensus under these conditions is maintained by the narrowing and purification of goals and methods. One manifestation of this process is the appearance of norms enjoining disinterest and the conscious rejection of external control; but its effects are more significant at a deeper level. Concern with narrowing and purification are reflected respectively in the high specificity and deliberate esotericity of scientific concepts. The fondness of scientists for Latin and Greek is more than a random eccentricity. Consensus is manifested in common acceptance of these special concepts and the rules and procedures that they specify; it is celebrated in the use of prescribed vocabulary and stylistic forms.

Some scientific terms are readily translated into everyday language and possess little independent meaning; the existence of a separate terminology in these cases allows for the future occurrence of independent conceptual development within and outside a discipline; meaning is never static; separate vocabulary ensures freedom from future social control. More often however everyday language and scientific terminology draw the boundaries of their discriminations in different places with different degrees of sharpness. Scientific concepts can achieve an impersonality and sharpness impossible with everyday terms because of the pragmatic role of the latter as social institutions.[12] Disputes about the

11. There is some evidence of tension between research indications and ethical constraints: see for example Pappworth (1967).

12. This argument does not imply a qualitative difference between scientific and everyday concepts. Key scientific concepts tend themselves to be imprecisely defined and to change in meaning over time: one can cite the physical notion of energy, the chemical notion of combination or the biological concept of the gene. Moreover, changes in meaning or in the sharpness of definition of such scientific concepts alter the way in which science is done. With their specialized objectives and narrow basis of cooperation, however, scientific disciplines provide a better setting for reaching agree-

meaning, and sharpness of definition, of such dichotomies as alive—dead, harmful—harmless, adulterated—improved, sane—insane or male—female often involve differences about how society ought to operate as well as matters for empirical investigation. By eschewing such concepts scientific disciplines avoid the divisive effects of external social conflicts.

Necessarily then scientific knowledge reaches the lay audience via a translation process. This can be straightforward: chemical knowledge of the stability and purity of substances can readily become advice about explosives or the nature of foodstuffs. But gearing science to problems couched in everyday pragmatic concepts can be extremely difficult; indeed in many cases evaluations deliberately excluded from a discipline's conceptual structure cannot be avoided if one is to so much as talk about a problem in everyday terms. Such a problem will not in general be able to generate a disciplinary consensus on how it is to be dealt with or even analysed, for scientific disciplines include actors with a wide variety of general social values and no institutionalized mechanisms exist for homogenizing these.

The concern of the sciences with purity and a limited area of relevance is, of course, common to all the professions and in marked contrast to the tendency of some knowledge sources to extend their scope and generalize their competence. In most professions the nature of agreed limitations and their effect on the bodies as knowledge sources can be clearly perceived from the outside; often they are formally articulated by the profession itself. Limitations of scientific competence, however, are rarely expressed[13] and difficult to describe simply; they become clear to the scientist as he uses his acquired paradigmatic framework. This conceptually built-in control is apparently as effective as any transmitted by rules and sanctions.

The favourable reception of scientific ideas by the lay public is

---

ment than society as a whole; the dynamic processes involved in establishing or altering sharply defined conceptual boundaries occur comparatively readily within a scientific discipline, where the effect upon an actor is circumscribed and the consequences for the institution predictable to an extent.

13. It is probably a valid generalization that 'embryonic' sciences make formal statements of their scope and objectives far more frequently than 'mature' ones.

thus, even in the context of general deference to the authority of science, limited by low intelligibility and the lack of institutionalized procedures consonant with the role of scientific disciplines as externally oriented knowledge sources. As the latter factor is linked to the high importance of internal consensus in science and the former would require far more than mass dissemination of a few simple rules of scientific method for its eradication, it would seem that closer articulation of science and society is likely to be produced, if at all, only from long-drawn-out social changes; perhaps the growth of intermediate mediating roles, controlled from within science, will be one such change.

References

AGASSI, J. (1963), 'Towards an historiography of science', *History and Theory*, no. 2.

BARBER, B. (1961), 'Resistance by scientists to scientific discovery', *Science*, vol. 134, no. 3479, 1 September, pp. 596–602.

BEN-DAVID, J. (1960), 'Roles and innovations in medicine', *Amer. J. Sociol.*, vol. 65. pp. 557–68.

BRIGHT, J. R. (1964), 'On resistance to technological innovation', in *Research Development and Technological Innovation*, Irwin.

GRAZIA, A. DE (ed.) (1966), *The Velikovsky Affair*, University Books, New York.

HANSON, N. (1958), *Patterns of Discovery*, Cambridge University Press.

HESSE, M. (1965), *Models and Analogies in Science*, Sheed & Ward.

HOMANS, G. (1961), *Social Behaviour: Its Elementary Forms*, Harcourt, Brace & World.

KUHN, T. S. (1957), *The Copernican Revolution*, Random House.

KUHN, T. S. (1962), *The Structure of Scientific Revolutions*, Chicago University Press.

LUKES, S. (1967), 'Some problems about rationality', *European J. Sociol.*, vol. 8, pp. 247–64.

MAUSNER, B., and MAUSNER, J. (1955), 'A study of the anti-scientific attitude', *Scientific American*, February.

MULKAY, M. (1969), 'Some aspects of cultural growth in the natural sciences', *Social Research*, vol. 6, no. 1, Spring.

PAPPWORTH, M. H. (1967), *Human Guinea Pigs*, Routledge & Kegan Paul.

PARSONS, T. (1951), *The Social System*, Free Press.

PEEL, J. D. Y. (1969), 'Understanding alien belief systems', *Brit. J. Sociol.*, March, pp. 69–84.

POLANYI, M. (1958), *Personal Knowledge*, Chicago University Press.

POLANYI, M. (1967), 'The growth of science in society', *Minerva*, vol. 5, pp. 533–545.

Public Health Service Publication, no. 1103, (1963), *Smoking and Health*, Washington.
REISER, S. J. (1966), 'Smoking and health: the congress and causality', in S. A. Lakoff (ed.), *Knowledge and Power*, Free Press.
ROGERS, E. M. (1962), *The Diffusion of Innovation*, Free Press.
SAPOLSKY, H. M. (1968), 'Science, voters and the fluoridation controversy', *Science*, vol. 162, 25 October, pp. 427–33.
TOULMIN, S. (1961), *Foresight and Understanding*, Hutchinson.
WINCH, P. (1958), *The Idea of a Social Science*, Routledge & Kegan Paul.

16 Charles E. Rosenberg

Scientific Theories and Social Thought

Excerpt from C. E. Rosenberg, 'Science and American social thought', in D. Van Tassel and M. G. Hall (eds.), *Science and Society in the United States*, Dorsey, 1966, pp. 137–84.

Most aspects of social thought deal with universal human problems, problems which must be answered by each generation in its own way, the need, for example, of society to arrive at an explanation of why some persons are wealthy, others poor, some healthy and vigorous, others sickly and weak. Throughout the nineteenth century the world of science presented a multitude of words and concepts with which such traditional concerns could be discussed. Late nineteenth-century science, for example, provided publicists and scholars of the day with the concepts of the Darwinian synthesis; they employed these ideas to explain social structure and social function. Social Darwinism[1] is only the most familiar among a number of such instances. At the same time, for example, metaphors originating in physics and electrophysiology were made to explain variation in human capacities. Neither folk wisdom nor the physician's clinical acumen doubted that human beings differed greatly in intelligence. They differed as well in their resistance to disease, in their keenness of perception and response to stimuli. For centuries, physicians had spoken of the concepts of vital force and of irritability in an effort to explain such innate differences.

By mid-nineteenth century, scientists had come to accept the electrical nature of the nervous impulse. In 1852, Helmholtz, the German physiologist and physicist, accomplished his successful measurement of the rate of nervous conductivity. Aside from the philosophical implications of this discovery, it seemed to make

1. The best general survey of Social Darwinism in this country (U S A) is still that by Hofstadter (1955). See also Persons (1950). There is some question as to how pervasive Social Darwinism as a well-articulated doctrine actually was. See Wyllie (1959).

almost inescapable the conclusions that nervous force might easily be the same as vital force and that this elusive vital force, if not electricity itself, must be some form of energy closely allied to it. In addition, the second law of thermodynamics[2] seemed to emphasize man's necessarily limited quantity of vital energy and the innumerable possibilities for energy loss from within the closed system that was the human organism.

The nervous system, as visualized by Americans before the acceptance of the neuron theory,[3] was a closed and continuous channel. A fixed quantity of nervous force, a hereditary endowment assumed to be electrical in nature, filled and coursed through this channel. (Popularizers soon grew fond of comparing the human brain and nervous system to the headquarters and wires of a great telegraph system. 'The brain is the central office, and in it there are nine hundred million cells generating nerve fluid. . . .'[4]) Using this schematic model, physicians and social thinkers were able to explain the most varied aspects of human behavior. Thus the artistically gifted maniac was the result of a particularly abnormal imbalance of nervous energies. 'The force which is turned away from some channel that is blocked up by disease rushes through the channels of sanity that remain unobstructed with heightened velocity.' Hypnotism or trance was the consequence of nervous force being concentrated in 'one direction'. Insanity might also be explained in these terms. No two persons would be born with the same amount of nervous force; no two persons would be subjected to the same external social pressures. Those individuals whose endowment of nervous force was inadequate to the pressures and crises of daily life would succumb to neurosis or psychosis.

2. The second law of thermodynamics states that in a free and continuous heat exchange, heat is always transferred from the hotter to the colder body.

3. This doctrine holds that the nervous system is composed of neurons or nerve cells. The neurons are structural units which are in contact with other units but not in continuity. The nervous pathways are conceived as chains of such units.

4. Hunter (c. 1900, p. 129). It was assumed in most of these analogies that man's endowment of nervous energy was not an absolute quantity provided at birth but rather the body's potential for the production of this nervous energy during life.

Such crude designs provoked little criticism in this self-consciously materialistic generation; electricity was not metaphysics and nervous force seemed far removed from vital force. We may smile today, perhaps, at the simplistic quality of these explanations, but it must be recalled that men as diverse as Sigmund Freud and Henry Adams were captivated by analogies in which energy relationships served to explain individual and collective behavior in a fashion not really different in form from these ingenuous schemes.[5]

This same mechanistic formulation helped not simply to explain mental illness and other extreme personal behavior; it served as well to express traditional social and moral sanctions in terms relevant to postbellum America. The electrical nature of the nerve impulse, the conservation of energy, and the second law of thermodynamics served as sources of didactic metaphors, metaphors clothed in the authority of science and yet dramatizing such long-standing concerns as the American desire to shore up middle-class morale and to provide a rationale for moderation in every aspect of behavior. These and other metaphors drawn from the sciences helped similarly to express the ambivalence of many Americans toward progress, toward urbanization, toward the treacherous fluidity of American life.

Moderation was, to nineteenth-century Americans, almost synonymous with morality. And excess was not only immoral, but, they believed, physiologically foolhardy as well. Man's limited complement of nervous energy, if considered in the light of the normal human need to indulge in a great variety of activities, meant that one should not indulge immoderately in any particular activity. Those which stimulated the emotions were particularly dangerous, for it seemed clear that nervous energy must necessarily be expended during the expression of strong emotions. The baleful effects of an excessive devotion to business might be the same as

5. The preceding quotations have been taken from Rosenberg (1962, p. 250). These formulations, it should be noted, preserved the traditional assumption that both heredity and environmental stress played a role in the etiology of nervous disease. There is, so far as I am aware, no study available of the metaphorical use of energy relationships in American social thought. For Freud, see Bernfeld (1944, pp. 341–62); for Henry Adams, see Jordy (1952). For an account of research relating to the discovery of electrical nature of the nervous impulse, see Liddell (1960, pp. 31–47.)

those of an obsessive love affair. All sensual pleasure was – in excess, of course – perilous. As Alexander Bain put it: 'Every throb of pleasure costs something to the physical system; and two throbs cost twice as much as one.'[6] Yet nervous energies must be discharged, for without physical or emotional release, these energies would accumulate and ultimately create pathological conditions. Sexual abstinence, for example, if not compensated for by a life of vigorous physical activity could lead to neurosis or, indeed, to any one of a score of ailments. Thus the frequency of hysteria in the more delicate and protected of middle-class maidens. Morality denied to 'young ladies' energy-discharging sexual activities, while fashion dictated their refraining from the active work or outdoor play which might also have reduced to a safe level the body's normal production of nervous energy.

Indeed, physicians argued, many aspects of American life made massive and unnatural demands on the nervous system. Constant choice and opportunity in business and religion, a lack of standards in personal and social life – all created tension and excitement. Competition began in the schools. The smallest of children were forced to specialize and to overwork in the quest for narrow academic distinction, their physical well-being forgotten. Even if the American survived his education with a minimum of psychic or physical damage, he had then to face an adulthood filled with insecurity. He lived his life at a pace too frenetic for relaxation or rest. American ways seemed especially alarming when compared with those of Europe. On the Continent, for example, religion was a source of security and passive reassurance; in the United States Protestantism made constant drafts upon moral and emotional reserves. In politics and business, too, any American, regardless of his social origin, might be subjected to the stress created by his own unbounded aspirations – and the consequent alternation of hope, of elation, of despair. 'The result of this extreme activity, is exhaustion and weakness. Physical bankruptcy is the result of drawing incessantly upon the reserve capital of nerve force.'[7]

6. 'The Correlation of Nervous and Mental Forces,' in Stewart (1875, p. 228). Bain, a professor at Aberdeen, was widely read in this country. Particularly significant in the attempt to popularize discoveries relating to the conservation of energy was the editorial work of Youmans. See Youmans (1869).

7. Pierce (1895, p. 619). For a clear exposition of the assumption that

More than a few Americans have felt misgivings in contemplating the dangerous and unsettling freedom of American life. Russian peasants chose neither their religion nor their rulers and suffered from none of the symptoms of nervous exhaustion, from wan faces and hollow cheeks, dyspepsia and sterility. 'Insanity', as one editorialist put it, 'is the skeleton at the feast of the highest civilization, . . . In proportion as nations have become free has mental disease multiplied.'[8] Physicians and social thinkers since the day of Thomas Jefferson and Benjamin Rush have criticized the peculiar tensions of American life. The speculative pathologies which explain precisely how these tensions injured the body have changed in form since the days of Rush, but the ambivalent attitudes which they express toward American life have not.

Yet neither Benjamin Rush nor his successors later in the century – S. Weir Mitchell, Isaac Ray, and G. M. Beard among others – were willing, warn as they might of the psychic perils of American life, to exchange its liberties for the placid tyranny of the Russian or Turkish empires (or, most Americans felt, their individualistic Protestantism for the formalistic reassurances of Catholicism). Throughout the nineteenth century, indeed, American warnings of the dangers of modern life to mental stability were not a negative but, in sum, a positive, even nationalistic, doctrine. Progress and liberty were unquestionably desirable, and the ailments which they induced in American minds were in a sense additional bits of evidence for the superiority of American ways of life. Technological change might be the cause of mental unease, but almost all Americans were relatively sanguine in their attitude toward the future of such material change. Many Americans believed that the very processes of technological change that appeared to threaten mental health would ultimately provide remedies. Such ills that might develop in the interval were those of a transitional period in history and a small price paid for social progress. 'We must not go backward, but forward,' Irving Fisher wrote in 1908. 'The cure for eye strain is not in ignoring the invention of reading, but in introducing the invention of glasses. The cure for

---

American life held dangers to mental health, see the many references in Dain (1964). See also Rosen (1959) and Altschule (1957, pp. 119–39).

8. *Independent* (New York), XXII (June 30, 1870), p. 4.

tuberculosis is not in the destruction of houses, but in devices for ventilation.'[9] Throughout the nineteenth century, American minds were marked by unresolved and ordinarily unstated contradictions – between an ingenuously arrogant nationalism and a chronic national insecurity, between optimism and pessimism, between primitivism and progress. With a peculiar appropriateness, science provided a vocabulary and a source of imagery in which these contradictions could be expressed with some subtlety – indeed with an unselfconscious sensitivity which might well have been unattainable by the same men in more formal modes of speculation.

A function central to the social thought of any place or time is the formulation and justification of an ideal social type. Until the present century, American formal rhetoric consistently pictured as ideal the responsible, middling, yeoman farmer. The virtues attributed to this thinking agriculturist were physical as well as moral. A rural upbringing seemed to guarantee a far healthier life than a youth spent in the debilitating air of the city. It seemed equally clear that a middling life was more wholesome physically – as well as far more likely to be wholesome morally – than that of either the very poor or the extremely rich. The lower classes, improperly fed, addicted to drink and indulging in other vices, living in improperly ventilated apartments, had little chance to live out a natural life span. The rich, whether they had inherited wealth or risen to it, were alike prey to nervous disease. The idle scion of wealthy parents was likely to suffer from the mental corrosion of inactivity if not from dissipation; the self-made man was necessarily exposed to the tensions and stress which accompanied his financial ambitions. The ideal citizen was rural, moderately prosperous, of good, though not brilliant mental endowment. Genius, authorities on heredity agreed, was often accompanied by insanity, or idiocy, or ill-health; it could hardly be held up as an ideal. Therefore, concluded a prominent eugenicist, expressing a long-felt American conviction, 'it would seem wise not to breed for geniuses but for a solid middle class'.[10]

9 Fisher (1909). See Rosenberg (1962, p. 257); Corning (1884, p. 135). For influential statements in regard to the influence of American life on mental health, see Ray (1863, especially pp. 219–23) and Rush (1812, pp. 65–69 and passim).

10. Davenport (1922) to D. Starr Jordan, January 27, 1922. The literature

And, he might have added, to raise them in contact with nature. Even the least credulous of eighteenth- and nineteenth-century physicians found it difficult to question the well-attested immunity of primitive peoples from most of the diseases which plagued civilized man. It seemed absurd to even suggest that a Congo Negro could suffer from neurasthenia, or a South Sea Islander from epilepsy. Heart disease, cancer, and even liver ailments, some authorities held, were never to be found in Non-Western peoples. Childbirth in the American Indian, travelers reported, was, if not completely painless, a relatively casual procedure. Even pauperism, it seemed natural to assume (Royce, 1880, vol. 1, p. 49),

> ... does not exist in the natural state of man. Under the sweet influences of the skies, he is in the woods as quick and nimble as the bird or deer he pursues. Only in the atmosphere, thick with moral and physical poison of crowded cities, he degenerates into a pauper, robbed of all that elasticity and high potency by which man masters every resistance....

Of course, those Americans who eulogized the health and virtue of primitive peoples would hardly have favored a return to such unsophisticated ways of life. Though it may have had its questionable aspects, Western civilization offered advantages superior to the health gained by that rude life of the savage. The crudeness of nervous organization which protected these simple peoples from illness, prevented them as well from creating a complex society. The highest of human activities – morality, religion, art, and literature – were all consequences of the more finely developed nervous organization of Western man. 'There can be no question', physicians assumed, 'as to whether the nervous systems of highly cultivated and refined individuals among civilized peoples are more complex and refined in structure and delicate in susceptibility and action, at least in their higher parts, than the nervous system of savages.'[11] It seemed clear enough that so delicate an apparatus

---

relating to the connection between genius and other forms of mental abnormality is immense. An excellent guide is Lange-Eichbaum (1961).

11. Jewell (1881). With the acceptance of Darwinism, it should be noted, these hypothetical mental attributes of civilized man were provided with evolutionary credentials. The traditionally higher faculties, the aesthetic and moral, were, as a matter of course identified with those highest in the scale of evolution. Being the 'highest', they were naturally man's last acqui-

could easily go awry. The sensitivity and complexity of this superior nervous organization, its greater area for the reception of sensation, its greater capacity for imagination, all helped explain the susceptibility of civilized man to nervous ailments.

There is a logic here, arbitrary and makeshift as these 'scientific' analogies may seem. It is to be found, not in their particular scientific content, but in their social function. We must look, not to the internal coherence of the scientific ideas appropriated, but at their external logic – that is, their social purpose. By way of illustration, let me refer to another example, this drawn from the field of legal medicine.

In many of the prominent 'insanity' trials held in the years between 1880 and 1900, medical witnesses for the prosecution and defense endorsed contradictory medical interpretations so as to bolster their legal position. Psychiatrists for the prosecution argued that the prisoner understood the nature and consequences of his act, appeared to reason coherently and, hence, following the generally accepted rule of law, was guilty. Psychiatrists for the defense, often held, on the other hand, that one might seem rational, even intelligent, and still not be responsible for one's actions. The cause of the individual's irresponsibility, they were convinced, lay frequently in heredity, in a congenital predisposition toward lack of moral perceptivity and control. Indeed, they argued, such men could be identified by physical stigmata which seemed characteristic of criminals.[12] Certainly such offenders

---

sition in his upward path. Primitive peoples were, quite literally, it was believed, more primitive, less complex in their cerebral development.

12. This doctrine is more commonly associated with the name and influence of Lombroso and his 'criminal anthropology', which sought to establish physical characteristics peculiar to the criminal type. Basic to this doctrine and preceding it in time was the theory of degeneration, a concept broadly influential throughout Western Europe in the second half of the nineteenth century. It held mental weakness and abnormality to be protean, hereditary, and progressive. Alcoholism might, in the next generation, appear as insanity, in the next, as idiocy. The basic study of degeneration is still that of Genil-Perrin, (1913). A valuable introduction to the influences of Lombrosianism in the United States, is that by Fink (1938, ch. 5–8). For a valuable contemporary survey which provides a good summary of an American's position, see Talbot (1899). Davis' study (1958), provides useful introductory material on earlier views of criminality.

could be recognized at autopsy, for the brain of a moral defective or habitual criminal was demonstrably different from that of a sane and law-abiding person.

Many people in the mid-twentieth century may well sympathize with these physicians and lawyers whom they see as fighting to save the mentally ill from capital punishment, applaud their citation of German authorities, their invocation of science, and opposition to the vengeful moralism of the prosecution's expert witnesses. Yet a closer examination of the content of their ideas reveals that the ideas of the prosecution, *not* the defense, experts are the ones, in form, at least, that are in some ways closer to modern views. The current tendency is to explain personality in dynamic terms and to reject the ideas that insanity is inevitably hereditary, that it manifests itself in gross pathological lesions, or that it is necessarily localized in one portion of the brain. Defenders of traditional morality at the end of the ninteenth century refused also to accept such materialistic arguments. The human personality was a whole, they argued, and one part of it, such as the moral sense, could not be diseased while another, the reasoning faculty, for example, remained healthy. They viewed the criminal act as the consequence not of a diseased or deformed brain but as a result of traits originating in the habitual actions of earliest childhood.[13]

In a sense, of course, this analysis is irrelevant; defense witnesses seem liberal today not because their ideas are correct in detail but because their values are acceptable, because it is more important that they quote German sources as transcendent authority – even if the authorities are wrong – rather than the Bible or the rules of criminal jurisprudence. The heart of the matter lies in one's attitude toward the criminal offender. Then, as now, forensic 'liberals' assumed a deterministic stance, 'conservatives' a less deterministic one. That the scientific ideas with which the determinism is justified have changed in this century seems less

13. This pattern was a common one, though, of course, only in the trials of prominent persons were a number of physicians recruited by both sides. In the most famous 'insanity' trial of the period, that of Charles J. Guiteau for the murder of President Garfield, the twenty-odd witnesses for the prosecution and defense divided in this fashion. Compare, for example, the testimony of E. C. Spitzka and Francis Kiernan for the defense with that of John P. Gray and Fordyce Barker for the prosecution. *Report of the Proceedings in the Case of the United States vs. Charles J. Guiteau* (1882).

important than that they serve the same social function. There is, then, a logic here, but a logic imposed by social necessity. It could not, in this case, be imposed by the content of the scientific arguments themselves; the factors determining the development of criminal behavior were – and in many ways still are – simply a subject of unverifiable speculation. The more tenuous an area of scientific knowledge, the smaller its verifiable content, the more easily its data may be bent to social purposes.

This is illustrated clearly by the manner, for example, in which ostensibly scientific formulations have found quite different social roles in different national contexts. A case in point is the differing reception accorded phrenology in the United States and in Europe. (The central notion of phrenology is that particular localized portions of the brain control the several aspects of human behavior. Phrenology was immensely popular in mid-nineteenth-century America.) American historians have seen phrenology as a widely popular, an optimistic and Anti-Calvinistic doctrine.[14] It promised understanding and control of man's personality. In Europe, however, phrenology did not parktake of these qualities; it assumed a far more deterministic cast. It seemed optimistic in this country simply because Americans would have it so. There is actually more reason logically to interpret phrenology, with its emphasis upon an anatomical basis for behavioral characteristics, as a pessimistic, deterministic doctrine. Yet Americans, eager as they were to find justifications for a hopeful view of human potential, could not ignore so plausible a scheme, one which promised a neat and attractively mechanistic explanation of human behavior. Americans could not ignore the doctrine, but they could – and did – quite easily ignore its more deterministic implications.

At this point a word of caution is due. The social historian, though he has no consuming interest in the history of scientific ideas as such, must still – if he is to investigate the role of these concepts in social thought – comprehend something of science as it was understood by men in the past. Without such knowledge he will inevitably be influenced by the assumptions of modern science.

14. A modern study of phrenology in the United States is that by Davis (1955). For Gall and the development of phrenology, see Temkin (1947), and Ackerknecht and Vallois (1956).

To the extent that this is the case, he will in some measure fail to assess the thought of the period he has elected to study.

This danger is illustrated clearly in the case of nineteenth-century attitudes toward heredity, especially in the use of heredity to explain disease and antisocial behavior. It is natural for us to associate a hereditarian emphasis with social conservatism and a comparatively pessimistic view of the potential for social reform. Yet this was not at all the case until roughly 1900 and the beginning of modern genetics. At no time in the nineteenth century did either physicians or laymen doubt that environmental changes (especially changes originating in long-standing habit or acute pathological conditions) could be transmitted from parents to their children.

Thus, for example, the significance of Richard Dugdale's study of the Jukes family conducted in the mid-1870s is quite different from that usually attributed to it. Dugdale, a self-consciously scientific reformer on many social fronts, was particularly interested in prison reform. As a member of the executive committee of New York's Prison Association, Dugdale conducted a study of county jails in the state and was struck by the frequency with which he encountered a certain family name. Dugdale expanded his study, gathering what data he could concerning the background of this family – christened Jukes for the sake of anonymity. He concluded that the congenital inadequacy of this family – its immorality, criminality, idiocy, and insanity – had cost the state of New York over a million and a quarter dollars. But to Dugdale and most of his contemporaries, these alarming results did not provide a brief for eugenic marriage laws or compulsory sterilization. They made, on the contrary, an urgent plea for environmental reform. That the Jukes's antisocial traits could be inherited dramatized the need for immediate reform of the conditions in which they lived; otherwise, drinking, narcotics addiction, poor moral and hygienic surroundings would not simply menace one generation but would contaminate as well all succeeding generations. Environment, as Dugdale (1910, p. 66) himself put it,

> is the ultimate controlling factor in determining careers, placing heredity itself as an organized result of invariable environment....
>
> If these conclusions are correct,

he reasoned, in terms of social policy,

then the whole question of the control of crime and pauperism becomes possible, within wide limits, if the necessary training can be made to reach over two or three generations.[15]

By the 1880s and early 1890s, however, this attitude had begun to change. Dugdale's optimistic use of hereditarian arguments to bolster his melioristic position had been transformed into a defensive and hostile emphasis on the deterministic aspects of heredity. But a few short decades after Dugdale's work, many students of heredity were calling for the sterilization of the unfit – and no longer with enthusiasm for environmental reform. And it must be recalled that such environmental reform was still logically implied; almost without exception physicians and biologists in the 1890s still assumed that at least some pathological conditions exerted a deleterious influence on heredity. It was not until well into the first decade of the twentieth century that Weismannism was generally accepted. Thus between 1850, let us say, and 1900, the consensus of science in regard to heredity had not changed; yet social thinkers had clearly moved in this period toward an emphasis upon the deterministic quality of heredity. They had selected those scientific plausibilities which fitted most conveniently into their social needs and presuppositions. Clearly, the nature of this selection and the changed social needs it implies demands explanation. It is equally clear that it is the task of the social historian to explain such change; it is a change, however, which would not be recognized without some consideration of the scientific context from which had been drawn the figures who dramatized this social debate.

The social use of scientific concepts is more than arbitrary. Ultimately, the content of the scientific building blocks employed in social thought does have a limiting effect on their employment – though not until the particular ideas utilized acquire a generally agreed-upon definition. Thus, the gradual rejection of the assumption that acquired characteristics could be inherited made inevitable a connection between thoroughgoing hereditarianism and

15. Despite the abundance of historical treatment accorded the evolution problem, there is, so far as I am aware, no comprehensive study of heredity in nineteenth-century thought.

social conservatism.[16] Though it took several decades, the acceptance of the new genetics created a situation in which hereditarianism became the stronghold of social conservatives; the reformists in temperament tended – as the logic of their emotional position dictated – to dissociate behavioral characteristics entirely from hereditary determination. They had no choice, for they too felt hereditary characteristics to be immutable. Yet their motivation was primarily humanitarian, and thus humanitarianism demanded the performance of those melioristic acts which social conservatives derided as useless in the face of implacable heredity.

References

ACKERKNECHT, E. H., and VALLOIS, H. (1956), *Franz Joseph Gall, Inventor of Phrenology and his Collection, University of Wisconsin Medical School,* University of Wisconsin Studies in Medical History, no. 1, Madison.

ALTSCHULE, M. D. (1957), *Roots of Modern Psychiatry,* New York.

BERNFELD, S. (1944), 'Freud's earliest theories and the school of Helmholtz', *Psychoanalyst Q.,* no. 13, pp. 341–62.

CORNING, L. (1884), *Brain Exhaustion, with Some Preliminary Considerations on Cerebral Dynamics,* New York.

DAIN, N. (1964), *Concepts of Insanity in the United States, 1798–1865,* Rutgers University Press.

DAVENPORT, C. B. (1922), Carnegie Institute of Washington, Department of Genetics, Cold Spring Harbor, New York.

DAVIS, D. B. (1958), *Homicide in American Fiction, 1798–1860* (Ithaca, 1958).

DAVIS, J. D. (1955), *Phrenology, Fad and Science, A 19th Century American Crusade,* New Haven.

DUGDALE, R. (1910), *The Jukes, A Study in Crime, Pauperism, Disease and Heredity,* New York and London, 4th edn.

FINK, A. E. (1938), *Causes of Crime. Biological Theories in the United States, 1880–1915,* Philadelphia.

FISHER, I. (1909), 'Bulletin 30 of the committee of one hundred on national health: being a report on national vitality, its wastes and conservation', Washington.

16. For a well-balanced, recent account of the eugenics movement, see Haller (1963). An illuminating contrast is that made by a comparative reading of Dugdale (1910) and Goddard (1912). Pastore's (1949) represents an attempt to correlate social views with attitudes toward heredity in a group of American scientists. For a study emphasizing the importance of Weismannism in the early development of the social sciences in this country, see Stocking (1962, pp. 239–56).

Genil-Perrin, G. P. H. (1913), *Histoire des Origines et de l'Evolution de l'Idée de Dégénérescence en Médecine Mentale*, Paris.
Goddard, H. H. (1912), *The Kallikak Family, A Study in the Heredity of Feeble-Mindedness*, New York.
Haller, M. H. (1963), *Eugenics Hereditarian Attitudes in American Thought*, Rutgers University Press.
Hofstadter, R. (1955), *Social Darwinism in American Thought*, Beacon Press.
Hunter, W. J. (c. 1900), *Manhood Wrecked and Rescued. How Strength or Vigor is lost, and How it May be Restored Through Self-Treatment*, Passaic, New Jersey.
Independent (1870), no. 22, New York, 30 June, p. 4.
Jewell, J. S. (1881), 'Influence of our present civilization in the production of nervous and mental disease', *J. Nervous & Mental Disease*, no. 8, p. 4.
Jordy, W. (1952), *Henry Adams: Scientific Historian*, New Haven.
Lange-Eichbaum, W. (1961), *Genie, Irrsinn und Ruhm*, Munich.
Liddell, E. G. T. (1960), *The Discovery of Reflexes*, Oxford.
Pastore, N. (1949), *The Nature-Nurture Controversy*, New York.
Persons, S. (ed.) (1950), *Evolutionary Thought in America*, New Haven.
Pierce, R. V. (1895), *The People's Common Sense Medical Adviser in Plain English, or Medicine Simplified*, Buffalo, New York.
Ray, I. (1863), *Mental Hygiene*, Boston.
Report of the Proceedings in the case of the United States *vs.* Charles J. Guiteau (1872), Washington.
Rosen, G. (1959), 'Social stress and mental disease from the eighteenth century to the present: some origins of social psychiatry', *Millbank Memorial Fund Q.*, no. 37, pp. 5–32.
Rosenberg, C. E. (1962), 'The place of George M. Beard in nineteenth-century psychiatry', *Bull. Hist. Med.*, vol. 86, pp. 245–59.
Royce, S. (1880), *Deterioration and the Elevation of Man through Race Education*, Boston.
Rush, B. (1812), *Medical Inquiries and Observations upon the Diseases of the Mind*, Philadelphia.
Stewart, B. (1875), *The Conservation of Energy*, New York.
Stocking, G. W., Jr (1962), 'Lamarckianism in American social science: 1890–1915', *J. Hist. Ideas.*, no. 23, pp. 239–56.
Talbot, E. S. (1899), *Degeneracy. Its Causes, Signs and Results*, London and New York.
Temkin, O. (1947), 'Gall and the phrenological movement', *Bull. Hist, of Med.*, no. 21, pp. 275–331.
Wyllie, I. G. (1959), 'Social Darwinism and the businessman', *Proc. Amer. Phil. Soc.*, no. 103, pp. 629–35.
Youmans, E. L. (1869), *Correlation and Conservation of Forces: A Series of Expositions, by Prof. Giove, Prof. Helmholtz, Dr Mayer, Dr Farraday, Prof. Liebig and Dr Carpenter*, New York.

Part Six
Scientific Concepts and the Nature of Society

In studying alien communities anthropologists have always felt it proper, indeed desirable, to relate theories about the natural world to the social organisation of the societies professing them. To a large extent beliefs have been assumed to be socially determined; anthropologists wishing to demonstrate their autonomy found themselves involved in special pleading. The corresponding beliefs of our own societies are those of science, and with respect to these the situation was reversed. Their autonomy was assumed, and sociologists gave little attention to their connection with other social factors. In practice, the presuppositions of the anthropologists produced a valuable research tradition where those of the sociologists justified inactivity; in strictly heuristic terms perhaps it is preferable that theories should over-connect?

The neglect of the sociology of scientific knowledge derived partially from factors internal to sociology. One was the tendency of sociologists to perceive science via ideal descriptions, which represented it as entirely the product of an absolutely valid general method – a tendency doubtless connected with the position of science as a professional ideal in sociology. The other was the tradition of establishing social determination in order to degrade the beliefs of others and expose them as erroneous – sociologism. More recently however, as these influences have declined, a transition has occurred so that now all areas of scientific activity – processes of discovery, social processes of validation and even rules of what are to count as observations and data, can respectably be examined from a sociological viewpoint; no effective taboos remain to limit the development of a fully general sociology of scientific knowledge-claims.[1] This transition is reviewed from a relativist standpoint in Dolby's contribution and linked with changes in the practice of the history and philosophy of science. The new relativism remains controversial, but it offers the prospect of a concrete research programme through which eventually it will doubtless come to be judged. And unlike older relativist orientations in the sociology of knowledge it involves no disparagement of the object of its study.

Among the few works based upon the possibility of the social determination of scientific knowledge, the most significant and provocative are those which relate scientific concepts and theories *in general* to social structure. Some of Durkheim's final work must, in the last analysis, fall into this category,[2] but it is represented here by contributions from Veblen and Marcuse, which are more explicitly related to modern science. No work of this nature can avoid epistemological implications, and the theories of knowledge of the three authors provide fascinating contrasts.

Unlike Veblen, Marcuse has had the good fortune to see his work inspiring the research of others. The paper by Habermas uses his conception of science as ideology to develop a broad view of the nature of present capitalist society and the place of science and technology within it. As one of the few attempts to characterise the general nature of modern 'scientific' society and its likely modes of development, it forms a fitting conclusion to this volume.

References

DURKHEIM, E., and MAUSS, M. (1963), *Primitive Classification*, trans. and intr. R. Needham, Routledge & Kegan Paul.

PARSONS, T. (1937), *The Structure of Social Action*, Free Press.

PEEL, J. D. Y. (1969), 'Understanding alien belief-systems', *Brit. J. Sociol.*, March, pp. 69–84.

1. Ideally, the sociologist who wishes to explain beliefs in general terms must maintain a position of externality to *all* beliefs, and concede none special status in advance of his investigations. (See Peel, 1969.) Formerly this was difficult with scientific beliefs, where attempts at sociological explanation aroused uniform 'a priori' opposition. Now 'sociological curiosity' can lead research where it will in this area.

2. See Durkheim and Mauss (1963). See also the discussion of Durkheim's epistemology in Parsons (1937).

17 R. G. A. Dolby

The Sociology of Knowledge in Natural Science

Excerpts from R. G. A. Dolby, 'The sociology of knowledge in natural science', *Science Studies*, vol. 1, no. 1, 1971.

This paper[1] is intended to explore the role of sociological factors in the knowledge-producing activity of natural science. The study of the production of scientific knowledge has always been a central issue in the history and philosophy of science. It is also a natural part of the sociological study of intellectual activities and their place in society, a separate pursuit which has arisen within this century. There has only been a partial interraction between these two separate approaches, and it is hoped that the present paper will promote cross-fertilization.

A difficult philosophical question lies behind much of the study of how social factors influence thought. This is the extent to which the validity of thought is dependent on the social influences in its genesis. In the spectrum of positions held over this issue, the prevailing theory of knowledge of natural science has had an important influence. Natural science has been seen as establishing its knowledge-claims by objective and absolute methods. This image has been generalized by some to all knowledge, and to the claim that there is an absolute independence of the process of establishing validity, from the social origins of belief. Others have used the same image and contrasted natural science with social thought, in which they have claimed that truth depends merely on consensus. The centre of such debates has been the status of social and political beliefs; the same scrutiny has not been applied so

1. This paper is a shortened version of one originally delivered at a conference on 'Problems and Prospects in the Sociology of Science', organized by the British Sociological Association Study Group on the Sociology of Science at Loughborough University of Technology, 3 January, 1970. The conference proceedings are to be published under the editorship of Irving Velody; the paper is reproduced in *Science Studies* by kind permission.

frequently to natural science. For sociologists have been heavily influenced by the heritage of the older nineteenth-century tradition of writings on scientific method. In their older form, these usually stressed the progressive and positive character of natural science, and the importance of an objective scientific method, in which the orderly collection of scientific facts was followed by their systematic expression in descriptive or theoretical laws. Such an account, by stressing objectivity in discovery as well as validation, seemed to leave no room for social factors in conditioning thought in natural science.

Sociology of knowledge in the first decades of this century studied natural science only to a small extent. This can be seen by reviewing the main influences in its emergence.[2] In Germany, sociology of knowledge was influenced primarily by developments of, and reactions to, Marxist ideas. This tended to make the centre of discussion the comparison of ideologies, and the method, the study of economic and other social factors in the knowledge-claims produced and accepted in a given social context.[3] In France, the main influence was the work of E. Durkheim and his school. Durkheim's influential studies found the origins of fundamental categories of human thought in the social organization of the tribe or community. (See for example, Durkheim and Mauss, 1903. Also Durkheim, 1923.) This again established problems and methods unrelated to modern natural science. In America, one of the early influences on the sociology of knowledge was the pragmatism of Peirce, James and Dewey. This was incorporated in, for example, T. Veblen's interesting work, around the turn of the century. But Veblen, while stressing the pragmatic origins of modern science, pointed out how its basic motivation of idle curiosity had allowed it to develop in a manner increasingly independent of its pragmatic social base. (See Veblen, 1961, p. 19.)

It is not surprising, then, that as the sociology of knowledge developed after the First World War, its concerns were only

2. A survey on these lines is given in the chapter on sociology of knowledge in Coser and Rosenberg (1964, pp. 667–84). This discussion is repeated by Coser (1968, vol. 8, pp. 428–35). A similar impression is given by Rybicki of Poland (1969).

3. Two of the later, more systematic exponents of this approach were M. Scheler and K. Mannheim, both of whom wrote a number of books on the sociology of knowledge.

marginally with natural science. But both the sociology of knowledge and the prevailing theories of methodology of science have changed since the beginning of this century. One of the most crucial developments in the theory of scientific method has been the rise of Logical Empiricism, especially with the Vienna Circle.[4] This retained the nineteenth-century conception of an objective scientific method, but stressed that it is a process of validation rather than discovery. Logical Empiricism has sought to relate scientific reasoning to the methods of modern mathematical logic. The scientist should justify his conclusions by reasoning from well established empirical premises expressed as statements based on particular observations, using methods of inference which are as explicit and as rigorous as possible. He may not reach certainty, but he should always strive for objectivity and rigour.

Logical Empiricism, as practised by philosophers of science such as R. Carnap, H. Reichenbach, C. G. Hempel and E. Nagel, together with such variants as the hypothetico-deductive method of K. R. Popper, has tended to dominate the field up to the early 1960s. And this work has influenced sociologists, both in the general questions of the methodology of sociology and in the issues of sociology of knowledge. One important distinction in this style of philosophy of science is the contrast made between the context of discovery and the context of justification.[5] This has provided a simple demarcation between philosophy of science and the sociology of scientific knowledge. The Logical Empiricist, who is only concerned with the process of validation of scientific results, can leave the question of the origin of scientific ideas to the psychologist and the sociologist. And the sociologist can accept the absolute nature of the justification process while asking questions about how scientists' beliefs arose.

The main developments in the sociology of knowledge have been associated with the transfer of dominance from Europe to America, and the development there of a different set of problems. In the introduction to his discussion of the sociology of knowledge

4. There is a large literature by and on the Vienna Circle. Most is referred to in the introduction, papers and bibliography of Ayer (1959).

5. This point is put quite forcefully by Logical Empiricists such as H. Reichenbach. See for example, Reichenbach (1958, p. 231). The same point is also made by Popper (1959, p. 31).

in *Social Theory and Social Structure*, Merton (1957, p. 439) contrasts the older European style of sociology of knowledge with the later American style. While the European discussion concentrated on what is of deepest significance, though amenable only to speculative investigation, the Americans have focused on what is measurable, even though it may not be especially significant. The Europeans have concentrated on systematized production of knowledge by élites, while the Americans studied the mass reception of unsystematized information.

In this American style of study, the problem area of sociology of knowledge has declined over the last two decades in favour of related areas of research such as the sociology of science, the sociology of mass communication and the sociology of professions. One factor in the general decline of sociology of knowledge might be the tendency, referred to by Merton, that this approach tends to lead people to ask questions about how people's beliefs arose, and to dismiss questions of validity. The sociologist of knowledge can avoid admitting that the views of someone else may be right. This was natural in the examination of diverse ideologies in Europe, but did not fit the American studies within a more uniform social system. American sociology of knowledge has usually followed Merton in accepting the Logical Empiricist distinction between discovery and validation.[6] Merton's influential 'Paradigm for the Sociology of Knowledge' (1957, pp. 460–7) systematically reviews the questions in the social origin of knowledge, setting aside the question of the validity of the knowledge-claims involved. The same distinction is accepted in the newer subject of sociology of science, which works within the framework of an absolute and objective methodology of validation in science.

Merton himself provides an example of a sociologist influenced

6. The general reluctance to relativize the theory of how scientific knowledge-claims are shown to be valid seems to come in part from the argument that a relativist theory of knowledge can only be justified on some absolute basis. I have attempted to avoid involving myself directly in this issue, but I should refer to Lavine's attempt (1942) to establish a theory of cognitive norms which could form the basis for such an unrestricted sociology of knowledge. Lavine argues against ten representative critics of the sociology of knowledge (mostly American) who agree that 'the sociological analysis of the elements of cognition be restrained from extension to "reason" or to the "logical schema of proof" or to "validity" ' (p. 342).

in many ways by the Logical Empiricist idea of scientific method. For example, in contrasting the method of historians and European sociologists of knowledge with the method of natural science, he makes these remarks (1957, p. 448):

> It is in fact, because effort is centred on successfully eliminating these differences of interpretation in science, because consensus is sought in place of diversity, that we can, with justification, speak of the cumulative nature of science. Among other things, cumulation requires reliability of initial observation. . . . works of science are as a matter of course placed one upon the others to comprise a structure of interlocking and mutually sustaining theories which permit an understanding of numerous observations. Towards this end, reliability of observation is of course a necessity.

The professionalization of history of science since the Second World War has been accompanied by a new understanding of scientific activity that has stimulated a radically divergent approach to Logical Empiricism in philosophy of science. This new approach promises to transform sociology of science in turn. The new movement is worth discussing in some detail in this context.

Ninteenth-century writings on scientific method claimed to be describing the process which successful scientists actually used, and which all scientists *ought* to use. There was no divergence between history and philosophy of science. But with the rise of Logical Empiricism, the descriptive claims of the nineteenth century seemed to be abandoned. With philosophers like R. Carnap, no attempt was made to reflect the activities of practising scientists. The explicit motive of the methodologist was to set out an ideal that scientists should *aspire* to follow. Other philosophers of science claimed greater descriptive validity for their work, but did not always clearly distinguish how much they saw it as prescriptive or descriptive. It is not difficult to demonstrate the separation between the logic of scientific method developed by Logical Empiricism and the actual scientific procedures revealed by historical study. Logical Empiricists, using mathematical logic as a model for natural science, seek a precision of expression, a rigour of argument and a minimalization of assumption that may be striven for but is never achieved in actual science, except in the pedagogical expression of the fields furthest from active research.

They are concerned with the methodology of validation of the assertions *produced* by science, while the scientist develops the techniques of discovery, the *process* of science.[7] They work with conceptual distinctions that seem only remotely related to the immediate concerns of the scientist – placing heavy stress on contrasts between particular observation and general law, between fact and theory. Popper's philosophy of science can be criticized in the same way. It, too, offers an ideal for which counter-examples from the practice of science are irrelevant. Philosophers with this approach use historical examples, but they tend to be idealized, neglecting many details of the historical evidence.

There would appear to be ample scope, then, for the rise of studies of the actual processes of scientific reasoning as they appear in historical practice. These should correct the excessively unrealistic idealizations of the Logical Empiricists. But the use of history of science cannot correct philosophy so easily. As Kuhn has argued,[8] people using historical material are very liable to distort it into a form fitting their philosophical preconceptions. It is very easy to select from historical material just those features that fit one's philosophical viewpoint, and very difficult to express the features that do not. This can be seen in the style of history of science that prevailed before the rise of professionalization of the field. Most history of science was written by scientists, especially scientists no longer active in reasearch. Such people had usually fully accepted the traditional image of scientific method. They saw themselves as chronicling the successive increments to the 'ever growing stockpile that constitutes scientific technique and knowledge'. (Kuhn, 1962, p. 2.) Their special concern was the assignment of credit for discovery, and condemnation of any who led the true path of science astray. The historiographical revolution that replaced this by a new style of history of science is discussed briefly by Kuhn (1962, ch. 1) and at greater length by J. Agassi (1963).[9]

7. This characterization of Logical Empiricism is expressed very neatly by the Logical Empiricist Rudner (1966, pp. 4–8). Rudner, however, does not admit how far his distinctions take him from the active concerns of the scientist.

8. In particular in a graduate seminar at Princeton University, in Spring 1966.

9. This gives a more extreme caricature of the old style, describing many

The new style of history of science claims to have something to tell the methodologist of science because it avoids the distortions of the philosopher's way of using historical evidence. The historian looks at his material in an attempt to fit as much of it as possible into a single coherent pattern. He does not select what he explains in terms of a preconceived pattern, but rather attempts to understand the historical processes so completely that he can explain all his evidence.[10] The most important lesson that the new style of history has learned is that the problems of scientists in other periods – the questions they asked, and the ways they tried to answer them – are often fundamentally different from the problems of modern science. If science is cumulative, it is also radically transformed in the process.

It is this historical insight (one which is not so new in the main stream of intellectual history[11] but which goes very much against the simple progressive cumulative image of natural science) which has encouraged the emergence of a new style of philosophy of science, closely related to history of science. Kuhn's *The Structure of Scientific Revolutions* is one of the better known works within this approach. A useful paper which gives a critical characterization of the approach is Shapere's (1966). Writers such as Hanson (1958), Feyerabend (1962, vol. 3, pp. 28–97 and 1965, pp. 145–260), Toulin (1961) and Kuhn have argued that theoretical and philosophical factors are presupposed in every aspect of scientific inquiry. Scientific method, as it is actually and inevitably carried out, loses its character of a logically straightforward process, once it is realized that such factors are presupposed in the meanings of observational and theoretical terms, in the characterization of the problems tackled by a science, and in what is to count as solutions to those problems. The approach introduces a new relativism into philosophy of science, one which opens up new vistas for the sociology of scientific knowledge. [...]

The presuppositions on which a scientific field is based can be

examples, with inevitable controversy and distortion. It is useful reading, however, for those who use the history of science but are not historians.

10. T. S. Kuhn, in a seminar, Princeton University, Spring 1966. See also the Epilegomena to Collingwood (1946).

11. Collingwood (1946) gives one of the most valuable accounts of its gradual emergence.

modified by historical change of the accepted theories in the field. As Kuhn and others have put it, the old theory and the new can be 'incommensurable'. Similarly, rival schools can form in a scientific field, particularly in its early stages: schools that share so few presuppositions that objective comparison seems impossible. Logical Empiricism and its variants treat observational and theoretical knowledge differently in this respect. Most philosophers of science fully recognize the uncertainty of theory, and may be prepared to admit that the theories currently held may at present be justified by psychological and sociological factors. Popper (1951, vol. 2, p. 217), for example, recognizes that

> ... there is no doubt that we are all suffering under our own system of prejudices ...; that we all take many things as self-evident, that we accept them uncritically and even with the naive and cocksure belief that criticism is quite unnecessary; and scientists are no exception to this rule, even though they may have superficially purged themselves from some of the superficial prejudices in their particular field.

But it is usually agreed that theories are dependent on the level of positive observational knowledge, at least for the process of testing. At this level it is assumed that objective impartial description is possible. The relativist theories of the new approach to philosophy of science challenge this assumption of an objective basis for scientific knowledge. *Every* level of science presupposes theoretical factors – which in turn can be influenced by all sorts of social and psychological distortions. If we accept that theories and hypotheses, the most uncertain elements in scientific inquiry, can come to prominence partly through the stimulation of institutional and general cultural factors, then these arguments lead us to recognize a sociological and cultural relativity in *all* scientific knowledge-claims. Such influences cannot be systematically exposed just by considering their possibility. It is only when different groups with different theoretical approaches to similar problems are exposed to one another in scientific debate, or in historical comparison, that contrasting presuppositions become clear. [...]

One of the clearest contrasts between scientific practice and the epistemological ideal of maximizing certainty is that science is essentially in the public domain. A scientist does not establish his own results. There can only be scientific knowledge of what a group of people can agree upon.[12] This immediately removes science from the subjective level of greatest certainty. It also introduces the possibility of relativism in the standards of those to whom a scientist directs his arguments. In considering objectivity in science, one must consider the audience for a knowledge-claim. It may be anticipated that the evaluation of a knowledge-claim will be affected by what the audience already accepts, what problems it is concerned with, what standards it applies, and what purposes the claim might be put to.

F. Znaniecki, in his book, *The Social Role of the Man of Science*, introduces the role of the 'social circle', the audience or public to which a thinker addresses himself. This circle imposes important influences on the thinking of the intellectual. In the logical ideals of scientific method, there is frequently no distinction between self-knowledge and arriving at a public consensus. But in fact, in arriving at self-conviction, all sorts of psychological and personal factors may operate. When a knowledge-claim is not addressed to an audience, these factors may never be eliminated, but such private assertions have hardly reached the status of scientific knowledge. There are strong social pressures on the scientist to address himself to a scientific audience, and it is only the rare misfit like Henry Cavendish (in his later electrical researches) who does not bother to report his discoveries. Now while there are many personal biases and distortions that play a role in scientific creativity, they do not assist in the persuasion of other scientists, who will clearly differ in their personal prejudices. Therefore, once the scientist starts attempting to convince other people, he must strive for objectivity, or at least for the avoidance of purely personal prejudices. It is like the detective who has convinced

12. This is the basis for Campbell's philosophy of science (see Campbell, 1920 and 1921). It is stressed in Ziman (1968). It is also an important part of the argument of Boring (1952). 'Scientific truth is thus usually *truth by agreement*, a social kind of truth.' It appears in the literature of the sociology of knowledge. For example, Child (1947–8), discusses the theoretical importance of a scientific consensus: 'Now consensus cannot, to be sure, establish the objective validity of a theory, but without an agreement of scientific opinion one cannot attain any surety at all about a proposition with a claim to scientific status. Such an agreement creates at least a high degree of probability in relation to the evidence; for scientific opinion results, by definition, from an assessment of the evidence critically and objectively.'

himself who the murderer is, but must now convince a jury in a court of law. When we discuss the audiences that the scientist may address himself to, we will see that the audience requirements made on his arguments vary. But the most important audience is that of fellow-specialists. For these, the scientist attempting to justify a knowledge-claim must relate his claims to the considerable amount of accepted knowledge in the field, using arguments that meet quite demanding standards. Particularly when he is publishing his work, he must conform to the standards of the field to have his work accepted as suitable for publication. He must make his work appear clear and conclusive.

It is in this sense that scientific work is, and must be, objective.[13] But while the scientific speciality demands high standards of presentation of evidence and arguments for conclusions, it also accepts far more as a basis for argument than some list of evident and uncontroversial facts. There are, as Kuhn argues, all sorts of procedural assumptions of strategy and technique which form part of the 'paradigm' of the community. Because such assumptions are not questioned, and because they can change historically, the objectivity is always relative to the particular specialist community. Objectivity is dependent on the consensus of a group of people who share a great number of tendentious assumptions. Michael Polanyi, in his discussions of the social nature of science, (see for example, 1946, 1951, 1958, 1967, 1969) has shown some of the factors that he finds present in scientific belief that contrast the consensus of a scientific community with the positivist ideal of objectivity. He points out the importance of tacit assumptions of plausibility and coherence in the climate of scientific opinion in which knowledge-claims are evaluated in a scientific field.

Such scientific communities are generally international and concerned with pure science, where the primary motive is the search for knowledge for its own sake. They are insulated from the prejudices of social groups most naturally treated in sociological study. But when the same people turn from pure science, as in the development of military products in war, their values and standards become more clearly identified with their social position. The

13. Lavine (1942, p. 354) discusses how this norm of objectivity in the sciences is in fact a way of refining the discrimination between personal idiosyncracy and 'community-centrism' (a term taken from Sherif, 1936).

behaviour of physical scientists in the Velikovsky affair reveals that scientists are not notably more objective than anybody else in dealing with issues that are not entirely within their own specialities. (See in particular, De Grazia, 1966. Also Polanyi, 1969.)

References

AGASSI, J. (1963), 'Towards an historiography of science', in *History and Theory*, Beiheft 2.

AYER, A. J. (1959), (ed.), *Logical Positivism*, Free Press.

BORING, E. G. (1952), 'The validation of scientific belief', *Proc. of the Amer. Philos. Soc.*, no. 96, pp. 535–90.

CAMPBELL, N. (1920), *Physics: the Elements*, Cambridge University Press

CAMPBELL, N. (1921), *What is Science?*, Cambridge University Press.

CHILD, A. (1947–8), 'The problem of truth in the sociology of knowledge', *Ethics*, no. 58, pp. 18–34.

COLLINGWOOD, R. G. (1946), The Idea of History, (ed.) Knox, Oxford University Press.

COSER, L. A. (1968), *International Encyclopedia of the Social Science.*

COSER, L. A., and ROSENBERG, B. (1946), *Sociological Theory: A Book of Readings*, Macmillan Co., 2nd edn.

DURKHEIM, E. (1923), *Formes Elémentaires de la Vie Religieuse*,

DURKHEIM, E., and MAUSS, M. (1903), 'De quelques formes primitives de classification', *L'Année Sociologique*, no. 6, pp. 1–72.

FEYERABEND, P. K. (1962), 'Explanation, reduction and empiricism', in *Minnesota Studies in the Philosophy of Science*, Minnesota.

FEYERABEND, P. K. (1965), 'Problems of empiricism', in R. G. Colodny (ed.), *Beyond the Edge of Certainty*, Prentice Hall.

GRAZIA, A. DE (1966) (ed.), *The Velikovsky Affair*, Sidgwick & Jackson

HANSON, N. (1958), *Patterns of Discovery*, Cambridge University Press.

KUHN, T. S. (1962), *The Structure of Scientific Revolution*, University of Chicago Press.

LAVINE, T. Z. (1942), 'Sociological analysis of cognitive norms', *J. of Philos.*, no. 39, pp. 342–56.

MERTON, R. K. (1957), *Social Theory and Social Structure*, Free Press, 1956.

POLANYI, M. (1946), *Science, Faith and Society*, University of Chicago Press, 1964.

POLANYI, M. (1951), *The Logic of Liberty*, University of Chicago Press.

POLANYI, M. (1958), *Personal Knowledge*, Harper & Row.

POLANYI, M. (1967a), *The Tacit Dimension*, Doubleday.

POLANYI, M. (1967b), 'The growth of science in society', *Minerva*, vol. 5, pp. 533–45.

POLANYI, M. (1969), *Knowing and Being*, University of Chicago Press.

POPPER, K. R. (1951), *The Open Society and its Enemies*, Routledge & Kegan Paul.

POPPER, K. R. (1959), *The Logic of Scientific Discovery*, Basic Books.
REICHENBACH, H. (1958), *The Rise of Scientific Philosophy*, University of California.
RUDNER, R. (1966), *Philosophy of Science*, Prentice-Hall.
RYBICKI, P. (1969), *International Encyclopedia of the Social Science*,
SHAPERE, D. (1966), 'Meaning and scientific change', in R. G. Colodny (ed.), *Mind and Cosmos: Essays in Contemporary Science and Philosophy*, Pittsburgh University Press.
SHERIF, M. (1936), *The Psychology of Social Norms*, Harper.
TOULMIN, S. (1961), *Foresight and Understanding*, Harper & Row.
VEBLEN, T. (1961), 'The place of science in modern civilization, (1906)', in *Science in Modern Civilization and other Essays*, Russell.
ZIMAN, J. (1968), *Public Knowledge: an Essay Concerning the Social Dimension of Science*, Cambridge University Press.

18 Thorstein Veblen

Idle Curiosity in Society

Excerpt from Thorstein Veblen, 'The place of science in modern Civilization', *American Journal of Sociology*, vol. 2, 1906, pp. 585–609. [References to the psychological literature have been omitted. Ed.]

The lower barbarian's knowledge of the phenomena of nature, in so far as they are made the subject of deliberate speculation and are organized into a consistent body, is of the nature of life-histories. This body of knowledge is in the main organized under the guidance of an idle curiosity. In so far as it is systematized under the canons of curiosity rather than of expediency, the test of truth applied throughout this body of barbarian knowledge is the test of dramatic consistency. In addition to their dramatic cosmology and folk legends, it is needless to say, these peoples have also a considerable body of worldly wisdom in a more or less systematic form. In this the test of validity is usefulness.[1]

The pragmatic knowledge of the early days differs scarcely at all in character from that of the maturest phases of culture. Its highest achievements in the direction of systematic formulation consist of didactic exhortations to thrift, prudence, equanimity, and shrewd management – a body of maxims of expedient conduct. In this field there is scarcely a degree of advance from Confucius to Samuel Smiles. Under the guidance of the idle curiosity, on the other hand, there has been a continued advance

1. 'Pragmatic' is here used in a more restricted sense than the distinctively pragmatic school of modern psychologists would commonly assign the term. 'Pragmatic', 'teleological', and the like terms have been extended to cover imputation of purpose as well as conversion to use. It is not intended to criticize this ambiguous use of terms, nor to correct it; but the terms are here used only in the latter sense, which alone belongs to them by force of early usage and etymology. 'Pragmatic' knowledge, therefore, is such as is designed to serve an expedient end for the knower, and is here contrasted with the imputation of expedient conduct to the facts observed. The reason for preserving this distinction is simply the present need of a simple term by which to mark the distinction between worldly wisdom and idle learning.

toward a more and more comprehensive system of knowledge. With the advance in intelligence and experience there come closer observation and more detailed analysis of facts. The dramatization of the sequence of phenomena may then fall into somewhat less personal, less anthropomorphic formulations of the processes observed; but at no stage of its growth – at least at no stage hitherto reached – does the output of this work of the idle curiosity lose its dramatic character. Comprehensive generalizations are made and cosmologies are built up, but always in dramatic form. General principles of explanation are settled on, which in the earlier days of theoretical speculation seem invariably to run back to the broad vital principle of generation. Procreation, birth, growth, and decay constitute the cycle of postulates within which the dramatized processes of natural phenomena run their course. Creation is procreation in these archaic theoretical systems, and causation is gestation and birth. The archaic cosmological schemes of Greece, India, Japan, China, Polynesia, and America, all run to the same general effect on this head. (See for example, Tylor, 1871, ch. 8.) The like seems true for the Elohistic elements in the Hebrew scriptures.

Throughout this biological speculation there is present, obscurely in the background, the tacit recognition of a material causation, such as conditions the vulgar operations of workday life from hour to hour. But this causal relation between vulgar work and product is vaguely taken for granted and not made a principle for comprehensive generalizations. It is overlooked as a trivial matter of course. The higher generalizations take their color from the broader features of the current scheme of life. The habits of thought that rule in the working-out of a system of knowledge are such as are fostered by the more impressive affairs of life, by the institutional structure under which the community lives. So long as the ruling institutions are those of blood-relationship, descent, and clannish discrimination, so long the canons of knowledge are of the same complexion.

When presently a transformation is made in the scheme of culture from peaceable life with sporadic predation to a settled scheme of predaceous life, involving mastery and servitude, gradations of privilege and honor, coercion and personal dependence, then the scheme of knowledge undergoes an analogous change. The predaceous, or higher barbarian, culture is, for the present purpose, peculiar in that it is ruled by an accentuated pragmatism. The institutions of this cultural phase are conventionalized relations of force and fraud. The questions of life are questions of expedient conduct as carried on under the current relations of mastery and subservience. The habitual distinctions are distinctions of personal force, advantage, precedence, and authority. A shrewd adaptation to this system of graded dignity and servitude becomes a matter of life and death, and men learn to think in these terms as ultimate and definitive. The system of knowledge, even in so far as its motives are of a dispassionate or idle kind, falls into the like terms, because such are the habits of thought and the standards of discrimination enforced by daily life.

The theoretical work of such a cultural era, as, for instance, the Middle Ages, still takes the general shape of dramatization, but the postulates of the dramaturgic theories and the tests of theoretic validity are no longer the same as before the scheme of graded servitude came to occupy the field. The canons which guide the work of the idle curiosity are no longer those of generation, blood-relationship, and homely life, but rather those of graded dignity, authenticity, and dependence. The higher generalizations take on a new complexion, it may be without formally discarding the older articles of belief. The cosmologies of these higher barbarians are cast in terms of a feudalistic hierarchy of agents and elements, and the causal nexus between phenomena is conceived animistically after the manner of sympathetic magic. The laws that are sought to be discovered in the natural universe are sought in terms of authoritative enactment. The relation in which the deity, or deities, are conceived to stand to facts is no longer the relation of progenitor, so much as that of suzerainty. Natural laws are corollaries under the arbitrary rules of status imposed on the natural universe by an all-powerful Providence with a view to the maintenence of his own prestige. The science that grows in such a spiritual environment is of the class represented by alchemy and astrology, in which the imputed degree of nobility and prepotency of the objects and the symbolic force of their names are looked to for an explanation of what takes place.

The theoretical output of the Schoolmen has necessarily an accentuated pragmatic complexion, since the whole cultural scheme under which they lived and worked was of a strenuously pragmatic character. The current concepts of things were then drawn in terms of expediency, personal force, exploit, prescriptive authority, and the like, and this range of concepts was by force of habit employed in the correlation of facts for purposes of knowledge even where no immediate practical use of the knowledge so gained was had in view. At the same time a very large proportion of the scholastic researches and speculations aimed directly at rules of expedient conduct, whether it took the form of a philosophy of life under temporal law and custom, or of a scheme of salvation under the decrees of an autocratic Providence. A naïve apprehension of the dictum that all knowledge is pragmatic would find more satisfactory corroboration in the intellectual output of scholasticism than in any system of knowledge of an older or a later date.

With the advent of modern times a change comes over the nature of the inquiries and formulations worked out under the guidance of the idle curiosity – which from this epoch is often spoken of as the scientific spirit. The change in question is closely correlated with an analogous change in institutions and habits of life, particularly with the changes which the modern era brings in industry and in the economic organization of society. It is doubtful whether the characteristic intellectual interests and teachings of the new era can properly be spoken of as less 'pragmatic,' as that term is sometimes understood, than those of the scholastic times; but they are of another kind, being conditioned by a different cultural and industrial situation.[2] In the life of the new era conceptions of authentic rank and differential dignity have grown weaker in practical affairs, and notions of preferential reality and authentic tradition similarly count for less in the new science. The forces at work in the external world are conceived in a less animistic manner, although anthropomorphism still prevails, at least to the degree required in order to give a dramatic interpretation of the sequence of phenomena.

2. As currently employed, the term 'pragmatic' is made to cover both conduct looking to the agent's preferential advantage, expedient conduct, and workmanship directed to the production of things that may or may not be of advantage to the agent. If the term be taken in the latter meaning, the culture of modern times is no less 'pragmatic' than that of the Middle Ages. It is here intended to be used in the former sense.

The changes in the cultural situation which seem to have had the most serious consequences for the methods and animus of scientific inquiry are those changes that took place in the field of industry. Industry in early modern times is a fact of relatively greater preponderance, more of a tone-giving factor, than it was under the régime of feudal status. It is the characteristic trait of the modern culture, very much as exploit and fealty were the characteristic cultural traits of the earlier time. This early-modern industry is, in an obvious and convincing degree, a matter of workmanship. The same has not been true in the same degree either before or since. The workman, more or less skilled and with more or less specialized efficiency, was the central figure in the cultural situation of the time; and so the concepts of the scientists came to be drawn in the image of the workman. The dramatizations of the sequence of external phenomena worked out under the impulse of the idle curiosity were then conceived in terms of workmanship. Workmanship gradually supplanted differential dignity as the authoritative canon of scientific truth, even on the higher levels of speculation and research. This, of course, amounts to saying in other words that the law of cause and effect was given the first place, as contrasted with dialectical consistency and authentic tradition. But this early-modern law of cause and effect – the law of efficient causes – is of anthropomorphic kind. 'Like causes produce like effects,' in much the same sense as the skilled workman's product is like the workman; 'nothing is found in the effect that was not contained in the cause,' in much the same manner.

These dicta are, of course, older than modern science, but it is only in the early days of modern science that they come to rule the field with an unquestioned sway and to push the higher grounds of dialectical validity to one side. They invade even the highest and most recondite fields of speculation, so that at the approach to the transition from the early-modern to the late-modern period, in the eighteenth century, they determine the outcome even in the counsels of the theologians. The deity, from having been in mediæval times primarily a suzerain concerned with the maintenance of his own prestige, becomes primarily a creator engaged in the workmanlike occupation of making things useful for man. His relation to man and the natural universe is no longer primarily that of a progenitor, as it is in the lower barbarian culture, but rather that of a talented mechanic. The 'natural laws' which the scientists of that era make so much of are no longer decrees of a preternatural legislative authority, but rather details of the workshop specifications handed down by the master-craftsman for the guidance of handicraftsmen working out his designs. In the eighteenth century science these natural laws are laws specifying the sequence of cause and effect, and will bear characterization as a dramatic interpretation of the activity of the causes at work, and these causes are conceived in a quasi-personal manner. In later modern times the formulations of causal sequence grow more impersonal and more objective, more matter-of-fact; but the imputation of activity to the observed objects never ceases, and even in the latest and maturest formulations of scientific research the dramatic tone is not wholly lost. The causes at work are conceived in a highly impersonal way, but hitherto no science (except ostensibly mathematics) had been content to do its theoretical work in terms of inert magnitude alone. Activity continues to be imputed to the phenomena with which science deals; and activity is, of course, not a fact of observation, but is imputed to the phenomena by the observer.[3] This is, also of course, denied by those who insist on a purely mathematical formulation of scientific theories, but the denial is maintained only at the cost of consistency. Those eminent authorities who speak for a colorless mathematical formulation invariably and necessarily fall back on the (essentially metaphysical) preconception of causation as soon as they go into the actual work of scientific inquiry.[4]

Since the machine technology has made great advances, during the nineteenth century, and has become a cultural force of wide-reaching consequence, the formulations of science have made another move in the direction of impersonal matter-of-fact. The machine process has displaced the workman as the archetype in whose image causation is conceived by the scientific investigators.

3. Epistemologically speaking, activity is imputed to phenomena for the purpose of organizing them into a dramatically consistent system.

4. See for example, Pearson (1892) and compare his ideal of inert magnitudes as set forth in his exposition with his actual work as shown in chaps. 9, 10 and 12, and more particularly in his discussions of 'Mother Right' and related topics in *The Chances of Death.* (1897).

The dramatic interpretation of natural phenomena has thereby become less anthropomorphic; it no longer constructs the life-history of a cause working to produce a given effect – after the manner of a skilled workman producing a piece of wrought goods – but it constructs the life-history of a process in which the distinction between cause and effect need scarcely be observed in an itemized and specific way, but in which the run of causation unfolds itself in an unbroken sequence of cumulative change. By contrast with the pragmatic formulations of worldly wisdom these latter-day theories of the scientists appear highly opaque, impersonal, and matter-of-fact; but taken by themselves they must be admitted still to show the constraint of the dramatic prepossessions that once guided the savage myth-makers.

In so far as touches the aims and the animus of scientific inquiry, as seen from the point of view of the scientists, it is a wholly fortuitous and insubstantial coincidence that much of the knowledge gained under machine-made canons of research can be turned to practical account. Much of this knowledge is useful, or may be made so, by applying it to the control of the processes in which natural forces are engaged. This employment of scientific knowledge for useful ends is technology, in the broad sense in which the term includes, besides the machine industry proper, such branches of practice as engineering, agriculture, medicine, sanitation, and economic reforms. The reason why scientific theories can be turned to account for these practical ends is not that these ends are included in the scope of scientific inquiry. These useful purposes lie outside the scientist's interest. It is not that he aims, or can aim, at technological improvements. His inquiry is as 'idle' as that of the Pueblo myth-maker. But the canons of validity under whose guidance he works are those imposed by the modern technology, through habituation to its requirements; and therefore his results are available for the technological purpose. His canons of validity are made for him by the cultural situation; they are habits of thought imposed on him by the scheme of life current in the community in which he lives; and under modern conditions this scheme of life is largely machine-made. In the modern culture, industry, industrial processes, and industrial products have progressively gained upon humanity, until these creations of man's ingenuity have latterly come to take the dominant place in the cultural scheme; and it is not too much to say that they have become the chief force in shaping men's daily life, and therefore the chief factor in shaping men's habits of thought. Hence men have learned to think in the terms in which the technological processes act. This is particularly true of those men who by virtue of a peculiarly strong susceptibility in this direction become addicted to that habit of matter-of-fact inquiry that constitutes scientific research.

Modern technology makes use of the same range of concepts, thinks in the same terms, and applies the same tests of validity as modern science. In both, the terms of standardization, validity, and finality are always terms of impersonal sequence, not terms of human nature or of preternatural agencies. Hence the easy copartnership between the two. Science and technology play into one another's hands. The processes of nature with which science deals and which technology turns to account, the sequence of changes in the external world, animate and inanimate, run in terms of brute causation, as do the theories of science. These processes take no thought of human expediency or inexpediency. To make use of them they must be taken as they are, opaque and unsympathetic. Technology, therefore, has come to proceed on an interpretation of these phenomena in mechanical terms, not in terms of imputed personality nor even of workmanship. Modern science, deriving its concepts from the same source, carries on its inquiries and states its conclusions in terms of the same objective character as those employed by the mechanical engineer.

So it has come about, through the progressive change of the ruling habits of thought in the community, that the theories of science have progressively diverged from the formulations of pragmatism, eversince the modern era set in. From an organization of knowledge on the basis of imputed personal or animistic propensity the theory has changed its base to an imputation of brute activity only, and this latter is conceived in an increasingly matter-of-fact manner; until, latterly, the pragmatic range of knowledge and the scientific are more widely out of touch than ever, differing not only in aim, but in matter as well. In both domains knowledge runs in terms of activity, but it is on the one hand knowledge of what had best be done, and on the other hand knowledge of what takes place; on the one hand knowledge of ways and means, on the other hand knowledge without any ulterior purpose. The latter range of knowledge may serve the ends of the former, but the converse does not hold true.

These two divergent ranges of inquiry are to be found together in all phases of human culture. What distinguishes the present phase is that the discrepancy between the two is now wider than ever before. The present is nowise distinguished above other cultural eras by any exceptional urgency or acumen in the search for pragmatic expedients. Neither is it safe to assert that the present excels all other civilizations in the volume or the workmanship of that body of knowledge that is to be credited to the idle curiosity. What distinguishes the present in these premises is (1) that the primacy in the cultural scheme has passed from pragmatism to a disinterested inquiry whose motive is idle curiosity, and (2) that in the domain of the latter the making of myths and legends in terms of imputed personality, as well as the construction of dialectical systems in terms of differential reality, has yielded the first place to the making of theories in terms of matter-of-fact sequence.

Pragmatism creates nothing but maxims of expedient conduct. Science creates nothing but theories. It knows nothing of policy or utility, of better or worse. None of all that is comprised in what is today accounted scientific knowledge. Wisdom and proficiency of the pragmatic sort does not contribute to the advance of a knowledge of fact. It has only an incidental bearing on scientific research, and its bearing is chiefly that of inhibition and misdirection. Wherever canons of expediency are intruded into or are attempted to be incorporated in the inquiry, the consequence is an unhappy one for science, however happy it may be for some other purpose extraneous to science. The mental attitude of worldly wisdom is at cross-purposes with the disinterested scientific spirit, and the pursuit of it induces an intellectual bias that is incompatible with scientific insight. Its intellectual output is a body of shrewd rules of conduct, in great part designed to take advantage of human infirmity. Its habitual terms of standardization and validity are terms of human nature, of human preference, prejudice, aspiration, endeavour, and disability, and the habit of mind that goes with it is such as is consonant with these terms. No doubt, the all-pervading pragmatic animus of the older and non-European civilizations has had more than anything else to do with their relatively slight and slow advance in scientific knowledge. In the modern scheme of knowledge it holds true, in a similar manner and with analogous effect, that training in divinity, in law, and in the related branches of diplomacy, business tactics, military affairs, and political theory, is alien to the skeptical scientific spirit and subversive of it.

References

PEARSON, K. (1892), *Grammar of Science*, London.
PEARSON, K. (1897), *The Chances of Death*, London.
TYLOR, E. B. (1871), *Primitive Culture*, London.

19 Herbert Marcuse

Technological Rationality and the Logic of Domination

Chapter 6 of Herbert Marcuse, *One-Dimensional Man*, Beacon Press, 1964.

In the social reality, despite all change, the domination of man by man is still the historical continuum that links pre-technological and technological Reason. However, the society which projects and undertakes the technological transformation of nature alters the base of domination by gradually replacing personal dependence (of the slave on the master, the serf on the lord of the manor, the lord on the donor of the fief, etc.) with dependence on the 'objective order of things' (on economic laws, the market, etc.). To be sure, the 'objective order of things' is itself the result of domination, but it is nevertheless true that domination now generates a higher rationality – that of a society which sustains its hierarchic structure while exploiting ever more efficiently the natural and mental resources, and distributing the benefits of this exploitation on an ever-larger scale. The limits of this rationality, and its sinister force, appear in the progressive enslavement of man by a productive apparatus which perpetuates the struggle for existence and extends it to a total international struggle which ruins the lives of those who build and use this apparatus.

At this stage, it becomes clear that something must be wrong with the rationality of the system itself. What is wrong is the way in which men have organized their societal labour. This is no longer in question at the present time when, on the one side, the great entrepreneurs themselves are willing to sacrifice the blessings of private enterprise and 'free' competition to the blessings of government orders and regulations, while, on the other side, socialist construction continues to proceed through progressive domination. However, the question cannot come to rest here. The wrong organization of society demands further explanation in view of the situation of *advanced* industrial society, in which the integration of the formerly negative and transcending social forces with the established system seems to create a new social structure.

This transformation of negative into positive opposition points up the problem: the 'wrong' organization, in becoming totalitarian on internal grounds, refutes the alternatives. Certainly it is quite natural, and does not seem to call for an explanation in depth, that the tangible benefits of the system are considered worth defending – especially in view of the repelling force of present day communism which appears to be the historical alternative. But it is natural only to a mode of thought and behaviour which is unwilling and perhaps even incapable of comprehending what is happening and why it is happening, a mode of thought and behaviour which is immune against any other than the established rationality. To the degree to which they correspond to the given reality, thought and behaviour express a false consciousness, responding to and contributing to the preservation of a false order of facts. And this false consciousness has become embodied in the prevailing technical apparatus which in turn reproduces it.

We live and die rationally and productively. We know that destruction is the price of progress as death is the price of life, that renunciation and toil are the prerequisites for gratification and joy, that business must go on, and that the alternatives are Utopian. This ideology belongs to the established societal apparatus; it is a requisite for its continuous functioning and part of its rationality.

However, the apparatus defeats its own purpose if its purpose is to create a humane existence on the basis of a humanized nature. And if this is not its purpose, its rationality is even more suspect. But it is also more logical, for from the beginning, the negative is in the positive, the inhuman in the humanization, enslavement in liberation. This dynamic is that of reality and not of the mind, but of a reality in which the scientific mind had a decisive part in joining theoretical and practical reason.

Society reproduced itself in a growing technical ensemble of things and relations which included the technical utilization of men – in other words, the struggle for existence and the exploitation of man and nature became ever more scientific and rational. The double meaning of 'rationalization' is relevant in this context. Scientific management and scientific division of labour vastly increased the productivity of the economic, political, and cultural enterprise. Result: the higher standard of living. At the same time and on the same ground, this rational enterprise produced a pattern of mind and behaviour which justified and absolved even the most destructive and oppressive features of the enterprise. Scientific-technical rationality and manipulation are welded together into new forms of social control. Can one rest content with the assumption that this unscientific outcome is the result of a specific societal *application* of science? I think that the general direction in which it came to be applied was inherent in pure science even where no practical purposes were intended, and that the point can be identified where theoretical Reason turns into social practice. In this attempt, I shall briefly recall the methodological origins of the new rationality, contrasting it with the features of the pre-technological model discussed in the previous chapter. [Not included here.]

The quantification of nature, which led to its explication in terms of mathematical structures, separated reality from all inherent ends and, consequently, separated the true from the good, science from ethics. No matter how science may now define the objectivity of nature and the interrelations among its parts, it cannot scientifically conceive it in terms of 'final causes'. And no matter how constitutive may be the role of the subject as point of observation, measurement, and calculation, this subject cannot play its scientific role as ethical or aesthetic or political agent. The tension between Reason on the one hand, and the needs and wants of the underlying population (which has been the object but rarely the subject of Reason) on the other, has been there from the beginning of philosophic and scientific thought. The 'nature of things,' including that of society, was so defined as to justify repression and even suppression as perfectly rational. True knowledge and reason demand domination over – if not liberation from – the senses. The union of Logos and Eros led already in Plato to the supremacy of Logos; in Aristotle, the relation between the god and the world moved by him is 'erotic' only in terms of analogy. Then the precarious ontological link between Logos and Eros is broken, and scientific rationality emerges as essentially neutral. What nature (including man) may be striving for is scientifically rational only in terms of the general laws of motion – physical, chemical, or biological.

Outside this rationality, one lives in a world of values, and values separated out from the objective reality become subjective. The only way to rescue some abstract and harmless validity for them seems to be a metaphysical sanction (divine and natural law). But such sanction is not verifiable and thus not really objective. Values may have a higher dignity (morally and spiritually), but they are not *real* and thus count less in the real business of life – the less so the higher they are elevated *above* reality.

The same de-realization affects all ideas which, by their very nature, cannot be verified by scientific method. No matter how much they may be recognized, respected, and sanctified, in their own right, they suffer from being non-objective. But precisely their lack of objectivity makes them into factors of social cohesion. Humanitarian, religious, and moral ideas are only 'ideal'; they don't disturb unduly the established way of life, and are not invalidated by the fact that they are contradicted by a behaviour dictated by the daily necessities of business and politics.

If the Good and the Beautiful, Peace and Justice cannot be derived either from ontological or scientific-rational conditions, they cannot logically claim universal validity and realization. In terms of scientific reason, they remain matters of preference, and no resuscitation of some kind of Aristotelian or Thomistic philosophy can save the situation, for it is *a priori* refuted by scientific reason. The unscientific character of these ideas fatally weakens the opposition to the established reality; the ideas become mere *ideals*, and their concrete, critical content evaporates into the ethical or metaphysical atmosphere.

Paradoxically, however, the objective world, left equipped only with quantifiable qualities, comes to be more and more dependent in its objectivity on the subject. This long process begins with the algebraization of geometry which replaces 'visible' geometric figures with purely mental operations. It finds its extreme form in some conceptions of contemporary scientific philosophy, according to which all matter of physical science tends to dissolve in mathematical or logical relations. The very notion of an objective substance, pitted against the subject, seems to disintegrate. From very different directions, scientists and philosophers of science arrive at similar hypotheses on the exclusion of particular sorts of entities.

For example, physics 'does not measure the objective qualities of the external and material world – these are only the results obtained by the accomplishment of such operations.' (Dingler, 1951, p. 360.) Objects continue to persist only as 'convenient intermediaries,' as obsolescent 'cultural posits'.[1] The density and opacity of things evaporate: the objective world loses its 'objectifiable' character, its opposition to the subject. Short of its interpretation in terms of Pythagorean-Platonic metaphysics, the mathematized Nature, the scientific reality appears to be ideational reality.

These are extreme statements, and they are rejected by more conservative interpretations, which insist that propositions in contemporary physics still refer to 'physical things'. (Grünbaum, 1954, pp. 85 *et seq.*) But the physical things turn out to be 'physical events,' and then the propositions refer to (and refer *only* to) attributes and relationships that characterize various kinds of physical things and processes. (Grünbaum, 1954, pp, 87 et seq.) Max Born states:

'... the theory of relativity ... has never abandoned all attempts to assign properties to matter ...' But 'often a measureable quantity is not a property of a thing, but a property of its *relation* to other things ... Most measurements in physics are not directly concerned with the things which interest us, but with some kind of projection, the word taken in the widest possible sense.' (Grünbaum, 1954, pp. 88 et seq.)[2]

And W. Heisenberg:

'Was wir mathematisch festlegen, ist nur zum kleinen Teil ein "objec-

1. Quine (1953, p. 44). Quine speaks of the 'myth of physical objects' and says that 'in point of epistemological footing the physical objects and the gods [of Homer] differ only in degree and not in kind' (*ibid.*). But the myth of physical objects is epistemologically superior 'in that it has proved more efficacious than other myths as a device for working a manageable structure into the flux of experience.' The evaluation of the scientific concept in terms of 'efficacious', 'device', and 'manageable' reveals its manipulative-technological elements.

2. Author's italics.

tives Faktum," zum grösseren Teil eine Uebersicht über Möglichkeiten.' (1948.)[3]

Now 'events', 'relations', 'projections', 'possibilities' can be meaningfully objective only for a subject – not only in terms of observability and measurability, but in terms of the very structure of the event of relationship. In other words, the subject here involved is a *constituting* one – that is, a possible subject for which some *data* must be, or can be conceivable as event or relation. If this is the case, Reichenbach's statement would still hold true: that propositions in physics can be formulated without reference to an *actual* observer, and the 'disturbance by means of observation,' is due, not to the human observer, but to the instrument as 'physical thing'. (Grünbaum, 1954, p. 85.)

To be sure, we may assume that the equations established by mathematical physics express (formulate) the actual constellation of atoms, i.e. the objective structure of matter. Regardless of any observing and measuring 'outside' subject A may 'include' B, 'precede' B, 'result in' B; B may be 'between' C, 'larger than' C, etc. – it would still be true that these relations imply location, distinction, and identity in the difference of A, B, C. They thus imply the capacity of *being* identical in difference, of *being* related to ... in a specific mode, of *being* resistant to other relations, etc. Only this capacity would be in matter itself, and then matter itself would be objectively of the structure of mind – an interpretation which contains a strong idealistic element:

... inanimate objects, without hesitation, without error, simply by their existence, are integrating the equations of which they know nothing. Subjectively, nature is not of the mind – she does not think in mathematical terms. But objectively, nature is of the mind – she can be thought in mathematical terms. (Weizsäcker, 1949, p. 20.)

A less idealistic interpretation is offered by Karl Popper (1957, pp. 155 *et seq.*), who holds that, in its historical development, physical science uncovers and defines different layers of one and the same objective reality.[4] In this process, the historically surpassed concepts are being cancelled and their intent is being

3. 'What we establish mathematically is "objective fact" only in small part, in larger part it is a survey of possibilities.'

4. Similarly: Bunge (1959, pp. 108 et seq.).

integrated into the succeeding ones – an interpretation which seems to imply progress toward the real core of reality, that is, the absolute truth. Or else reality may turn out to be an onion without a core, and the very concept of scientific truth may be in jeopardy.

I do not suggest that the philosophy of contemporary physics denies or even questions the reality of the external world but that, in one way or another, it suspends judgement on what reality itself may be, or considers the very question meaningless and unanswerable. Made into a methodological principle, this suspension has a twofold consequence: (a) it strengthens the shift of theoretical emphasis from the metaphysical 'What is . . . ?' (τί ἐστίν) to the functional 'How . . . ?', and (b) it establishes a practical (though by no means absolute) certainty which, in its operations with matter, is with good conscience free from commitment to any substance outside the operational context. In other words, theoretically, the transformation of man and nature has no other objective limits than those offered by the brute factuality of matter, its still unmastered resistance to knowledge and control. To the degree to which this conception becomes applicable and effective in reality, the latter is approached as a (hypothetical) system of instrumentalities; the metaphysical 'being-as-such' gives way to 'being-instrument'. Moreover, proved in its effectiveness, this conception works as an a *priori* – it predetermines experience, it *projects* the direction of the transformation of nature, it organizes the whole.

We just saw that contemporary philosophy of science seemed to be struggling with an idealistic element and, in its extreme formulations, moving dangerously close to an idealistic concept of nature. However, the new mode of thought again puts idealism 'on its feet'. Hegel epitomized the idealistic ontology: If Reason is the common denominator of subject and object, it is so as the synthesis of *opposites*. With this idea, ontology comprehended the *tension* between subject and object; it was saturated with concreteness. The reality of Reason was the playing out of this tension in nature, history, philosophy. Even the most extremely monistic system thus maintained the idea of a substance which unfolds itself in subject and object – the idea of an antagonistic reality. The scientific spirit has increasingly weakened this antagonism. Modern scientific philosophy may well begin with the notion of two substances, *res cogitans* and *res extensa* – but as the extended matter becomes comprehensible in mathematical equations which, translated into technology, 'remake' this matter, the *res extensa* loses its character as independent substance.

The old division of the world into objective processes in space and time and the mind in which these processes are mirrored – in other words, the Cartesian difference between *res cogitans* and *res extensa* – is no longer a suitable starting point for our understanding of modern science. (Heisenberg, 1958, p. 29.)[5]

The Cartesian division of the world has also been questioned on its own grounds. Husserl pointed out that the Cartesian *Ego* was, in the last analysis, not really an independent substance but rather the 'residue' or limit of quantification; it seems that Galileo's idea of the world as a 'universal and absolutely pure' *res extensa* dominated *a priori* the Cartesian conception. (See Biemel, 1954, p. 81.) In which case the Cartesian dualism would be deceptive, and Descartes' thinking ego-substance would be akin to the *res extensa*, anticipating the scientific subject of quantifiable observation and measurement. Descartes' dualism would already imply its negation; it would clear rather than block the road toward the establishment of a one-dimensional scientific universe in which nature is 'objectively of the mind', that is, of the subject. And this subject is related to its world in a very special way:

'... la nature est mise sous le signe de l'homme actif, de l'homme inscrivant la technique dans la nature.'[6]

The science of nature develops under the *technological a priori* which projects nature as potential instrumentality, stuff of control and organization. And the apprehension of nature as (hypothetical) instrumentality *precedes* the development of all particular technical organization:

5. Heisenberg writes: 'The "thing-in-itself" is for the atomic physicist, if he uses this concept at all, finally a mathematical structure; but this structure is – contrary to Kant – indirectly deduced from experience.' (1959, p. 83).

6. 'Nature is placed under the sign of active man, of the man who inscribes technique in nature.' Bachelard (1951, p. 7), with reference to Marx and Engels (1846, p. 163, et seq.).

Modern man takes the entirety of Being as raw material for production and subjects the entirety of the object-world to the sweep and order of production (*Herstellen*). . . the use of machinery and the production of machines is not technics itself but merely an adequate instrument for the realization (*Einrichtung*) of the essence of technics in its objective raw material. (Heidegger, 1950, pp. 266 et seq. Also 1954, pp. 22, 29.)[7]

The technological *a priori* is a political *a priori* inasmuch as the transformation of nature involves that of man, and inasmuch as the 'man-made creations' issue from and re-enter a societal ensemble. One may still insist that the machinery of the technological universe is 'as such' indifferent towards political ends – it can revolutionize or retard a society. An electronic computer can serve equally a capitalist or socialist administration; a cyclotron can be an equally efficient tool for a war party or a peace party. This neutrality is contested in Marx's controversial statement that the 'hand-mill gives you society with the feudal lord: the steam-mill society with the industrial capitalist'. (1935, p. 355.) And this statement is further modified in Marxian theory itself: the social mode of production, not technics is the basic historical factor. However, when technics becomes the universal form of material production, it circumscribes an entire culture; it projects a historical totality – a 'world'.

Can we say that the evolution of scientific method merely 'reflects' the transformation of natural into technical reality in the process of industrial civilization? To formulate the relation between science and society in this way is assuming two separate realms and events that meet each other, namely, (1) science and scientific thought, with their internal concepts and their internal truth, and (2) the use and application of science in the social reality. In other words, no matter how close the connection between the two developments may be, they do not imply and define each other. Pure science is not applied science; it retains its identity and validity apart from its utilization. Moreover, this notion of the essential *neutrality* of science is also extended to technics. The machine is indifferent toward the social uses to which it is put, provided those uses remain within its technical capabilities.

In view of the internal instrumentalist character of scientific

7. Translated by the author.

method, this interpretation appears inadequate. A closer relationship seems to prevail between scientific thought and its application, between the universe of scientific discourse and that of ordinary discourse and behaviour – a relationship in which both move under the same logic and rationality of domination.

In a paradoxical development, the scientific efforts to establish the rigid objectivity of nature led to an increasing de-materialization of nature.

'The idea of infinite nature existing as such, this idea that we have to give up, is the myth of modern science. Science has started out by destroying the myth of the Middle Ages. And now science is forced by its own consistency to realize that it has merely raised another myth instead.' (Weizsäcker, 1949, p. 71.)

The process which begins with the elimination of independent substances and final causes arrives at the ideation of objectivity. But it is a very specific ideation, in which the object constitutes itself in a quite *practical* relation to the subject:

'And what is matter? In atomic physics, matter is defined by its possible reactions to human experiments, and by the mathematical – that is, intellectual – laws it obeys. We are *defining* matter as a possible object of man's manipulation.' (p. 142.)[8]

And if this is the case, then science has become in itself technological:

'Pragmatic science has the view of nature that is fitting for a technical age.' (p. 71.)

To the degree to which this operationalism becomes the centre of the scientific enterprise, rationality assumes the form of methodical construction; organization and handling of matter as the mere stuff of control, as instrumentality which lends itself to all purposes and ends – instrumentality *per se*, 'in itself.'

The 'correct' attitude toward instrumentality is the *technical* approach, the correct logos is *techno-logy*, which projects and responds to a *technological reality*.[9] In this reality, matter as well

8. Author's emphasis.

9. I hope I will not be misunderstood as suggesting that the concepts of mathematical physics are designed as 'tools', that they have a technical, practical intent. Techno-logical is rather the *a priori* 'intuition' or apprehension of the universe in which science moves, in which it constitutes itself

as science is 'neutral'; objectivity has neither a telos in itself nor is it structured towards a telos. But it is precisely its neutral character which relates objectivity to a specific historical Subject – namely, to the consciousness that prevails in the society by which and for which this neutrality is established. It operates in the very abstractions which constitute the new rationality – as an internal rather than external factor. Pure and applied operationalism, theoretical and practical reason, the scientific and the business enterprise execute the reduction of secondary to primary qualities, quantification and abstraction from 'particular sorts of entities'.

True, the rationality of pure science is value-free and does not stipulate any practical ends, it is 'neutral' to any extraneous values that may be imposed upon it. But this neutrality is a *positive* character. Scientific rationality makes for a specific societal organization precisely because it projects mere form (or mere matter – here, the otherwise opposite terms converge) which can be bent to practically all ends. Formalization and functionalization are, *prior* to all application, the 'pure form' of a concrete societal practice. While science freed nature from inherent ends and stripped matter of all but quantifiable qualities, society freed men from the 'natural' hierarchy of personal dependence and related them to each other in accordance with quantifiable qualities – namely, as units of abstract labour power, calculable in units of time. 'By virtue of the rationalization of the modes of labour, the elimination of qualities is transferred from the universe of science to that of daily experience.'[10] (Horkheimer and Adorno, 1947, p. 50.)

Between the two processes of scientific and societal quantification, is there parallelism and causation, or is their connection simply the work of sociological hindsight? The preceding discussion proposed that the new scientific rationality was in itself, in its very abstractness and purity, operational inasmuch as it developed under an instrumentalist horizon. Observation and experiment, the methodical organization and coordination of

---

as *pure* science. Pure science remains committed to the *a priori* from which it abstracts. It might be clearer to speak of the instrumentalist *horizon* of mathematical physics. See Bachelard (1958, p. 31).

10. Translated by the author.

data, propositions, and conclusions never proceed in an unstructured, neutral, theoretical space. The project of cognition involves operations on objects, or abstractions from objects which occur in a given universe of discourse and action. Science observes, calculates, and theorizes from a position in this universe. The stars which Galileo observed were the same in classical antiquity, but the different universe of discourse and action – in short, the different social reality – opened the new direction and range of observation, and the possibilities of ordering the observed data. I am not concerned here with the historical relation between scientific and societal rationality in the beginning of the modern period. It is my purpose to demonstrate the *internal* instrumentalist character of this scientific rationality by virtue of which it is *a priori* technology, and the *a priori* of a *specific* technology – namely, technology as form of social control and domination.

Modern scientific thought, inasmuch as it is pure, does not project particular practical goals nor particular forms of domination. However, there is no such thing as domination *per se*. As theory proceeds, it abstracts from or rejects, a factual teleological context – that of the given, concrete universe of discourse and action. It is within this universe itself that the scientific project occurs or does not occur, that theory conceives or does not conceive the possible alternatives, that its hypotheses subvert or extend the pre-established reality.

The principles of modern science were *a priori* structured in such a way that they could serve as conceptual instruments for a universe of self-propelling, productive control; theoretical operationalism came to correspond to practical operationalism. The scientific method which led to the ever-more-effective domination of nature thus came to provide the pure concepts as well as the instrumentalities for the ever-more-effective domination of man by man *through* the domination of nature. Theoretical reason, remaining pure and neutral, entered into the service of practical reason. The merger proved beneficial to both. Today, domination perpetuates and extends itself not only through technology but *as* technology, and the latter provides the great legitimation of the expanding political power, which absorbs all spheres of culture.

In this universe, technology also provides the great rationalization of the unfreedom of man and demonstrates the 'technical' impossibility of being autonomous, of determining one's own life. For this unfreedom appears neither as irrational nor as political, but rather as submission to the technical apparatus which enlarges the comforts of life and increases the productivity of labour. Technological rationality thus protects rather than cancels the legitimacy of domination, and the instrumentalist horizon of reason opens on a rationally totalitarian society:

'On pourrait nommer philosophie autocratique des techniques celle qui prend l'ensemble technique comme un lieu où l'on utilise les machines pour obtenir de la puissance. La machine est seulement un moyen; la fin est la conquête de la nature, la domestication des forces naturelles au moyen d'un premier asservissement: la machine est un esclave qui sert à faire d'autres esclaves. Une pareille inspiration dominatrice et esclavagiste peut se rencontrer avec une requête de liberté pour l'homme. Mais il est difficile de se libérer en transférant l'esclavage sur d'autres êtres, hommes, animaux ou machines; régner sur un peuple de machines asservissant le monde entier, c'est encore régner, et tout règne suppose l'acceptation des schèmes d'asservissement.'[11]

The incessant dynamic of technical progress has become permeated with political content, and the Logos of technics has been made into the Logos of continued servitude. The liberating force of technology – the instrumentalization of things – turns into a fetter of liberation; the instrumentalization of man.

This interpretation would tie the scientific project (method and theory), *prior* to all application and utilization, to a specific societal project, and would see the tie precisely in the inner form of scientific rationality, i.e., in the functional character of its concepts. In other words, the scientific universe (that is, not the specific propositions on the structure of matter, energy, their

11. 'One might call autocratic a philosophy of technics which takes the technical whole as a place where machines are used to obtain power. The machine is only a means; the end is the conquest of nature, the domestication of natural forces through a primary enslavement: the machine is a slave which serves to make other slaves. Such a domineering and enslaving drive may go together with the quest for human freedom. But it is difficult to liberate oneself by transferring slavery to other beings, men, animals, or machines; to rule over a population of machines subjecting the whole world means still to rule, and all rule implies acceptance of schemata of subjection.' Simondon (1958, p. 127).

interrelation, etc., but the projection of nature as quantifiable matter, as guiding the hypothetical approach to – and the mathematical-logical expression of – objectivity) would be the horizon of a concrete societal practice which would be *preserved* in the development of the scientific project.

But, even granting the internal instrumentalism of scientific rationality, this assumption would not yet establish the *sociological* validity of the scientific project. Granted that the formation of the most abstract scientific concepts still preserves the interrelation between subject and object in a given universe of discourse and action, the link between theoretical and practical reasons can be understood in quite different ways.

Such a different interpretation is offered by Jean Piaget in his 'genetic epistemology'. Piaget interprets the formation of scientific concepts in terms of different abstractions from a general interrelation between subject and object. Abstraction proceeds neither from the mere object, so that the subject functions only as the neutral point of observation and measurement, nor from the subject as the vehicle of pure cognitive Reason. Piaget (1950, p. 287) distinguishes between the process of cognition in mathematics and in physics. The former is abstraction 'à l'intérieur de l'action comme telle':

'Contrairement à ce que l'on dit souvent, les êtres mathématiques ne résultent donc pas d'une abstraction à partir des objects, mais bien d'une abstraction effectuée au sein des actions comme telles. Réunir, ordonner, déplacer, etc. sont des actions plus générales que penser, pousser, etc., parce qu'elles tiennent à la coordination même de toutes les actions particulières et entrent en chacune d'elles à titre de facteur coordinateur . . .'[12]

Mathematical propositions thus express 'une accomodation générale à l'objet' – in contrast to the particular adaptations which are characteristic of true propositions in physics. Logic and mathematical logic are 'une action sur l'objet quelconque, c'est-

12. 'Contrary to what is often said, mathematical entities are not therefore the result of an abstraction based on objects but rather of an abstraction made in the midst of actions as such. To assemble, to order, to move, etc., are more general actions than to think, to push, etc., because they insist on the coordination itself of all particular actions and because they enter into each of them as coordinating factor.'

à-dire une action accomodée de façon générale' (p. 288); and this 'action' is of general validity inasmuch as

'cette abstraction ou différenciation porte jusqu'au sein des coordinations héréditaires, puisque les mécanismes coordinateurs de l'action tiennent toujours, en leur source, à des coordinations réflexes et instinctives.' (p. 289.)[13]

In physics, abstraction proceeds from the object but is due to specific actions on the part of the subject, thus abstraction assumes necessarily a logico-mathematical form because

'des actions particulières ne donnent lieu à une connaissance que coordonnées entre elles et que cette coordination est, par sa nature même, logico-mathématique.' (p. 291.)[14]

Abstraction in physics leads necessarily back to logico-mathematical abstraction and the latter is, as pure coordination, the general form of action – 'action as such' ('*l'action comme telle*'). And this coordination constitutes objectivity because it retains hereditary, 'reflexive and instinctive' structures.

Piaget's interpretation recognizes the internal practical character of theoretical reason, but derives it from a general structure of action which, in the last analysis, is a hereditary, biological structure. Scientific method would ultimately rest on a biological foundation, which is supra- (or rather infra-) historical. Moreover, granted that all scientific knowledge presupposes coordination of particular actions, I do not see why such coordination is 'by its very nature' logico-mathematical – unless the 'particular actions' are the scientific operations of modern physics, in which case the interpretation would be circular.

In contrast to Piaget's rather psychological and biological analysis, Husserl (see Biemel, ed., 1954) has offered a genetic epistemology which is focussed on the socio-historical structure of scientific reason. I shall here refer to Husserl's work only insofar as it em-

13. 'This abstraction or differentiation extends to the very centre of hereditary coordinations because the coordinating mechanisms of the action are always attached, at their source, to coordinations by reflex and instinct.'

14. 'Particular actions result only in knowledge if they are coordinated among them and if this coordination is in its very nature logical-mathematical.'

phasizes the extent to which modern science is the 'methodology' of a pre-given historical reality within whose universe it moves.

Husserl starts with the fact that the mathematization of nature resulted in valid practical knowledge: in the construction of an 'ideational' reality which could be effectively 'correlated' with the *empirical* reality (pp. 19, 42). But the scientific achievement referred back to a *pre*-scientific practice, which constituted the original basis (the *Sinnesfundament*) of Galilean science. This pre-scientific basis of science in the world of practice (*Lebenswelt*), which determined the theoretical structure, was not questioned by Galileo; moreover, it was concealed (*verdeckt*) by the further development of science. The result was the illusion that the mathematization of nature created an 'autonomous (*eigenständige*) absolute truth' (p. 49, *et seq.*), while in reality, it remained a specific method and technique for the *Lebenswelt*. The ideational *veil* (*Ideenkleid*) of mathematical science is thus a veil of *symbols* which represents and at the same time masks (*vertritt* and *verkleidet*) the world of practice (p. 52).

What is the original, pre-scientific intent and content that is preserved in the conceptual structure of science? *Measurement* in practice discovers the possibility of using certain basic forms, shapes, and relations, which are universally 'available as identically the same, for exactly determining and calculating empirical objects and relations' (p. 25). Through all abstraction and generalization, scientific method retains (and masks) its pre-scientific-technical structure; the development of the former represents (and masks) the development of the latter. Thus classical geometry 'idealizes' the practice of surveying and measuring the land (*Feldmesskunst*). Geometry is the theory of practical objectification.

To be sure, algebra and mathematical logic construct an absolute ideational reality, freed from the incalculable uncertainties and particularities of the *Lebenswelt* and of the subjects living in it. However, this ideational construction *is* the theory and technic of 'idealizing' the new *Lebenswelt*:

'In the mathematical practice, we attain what is denied to us in the empirical practice, i.e., *exactness*. For it is possible to determine the ideal forms in terms of *absolute identity* . . . As such, they become universally available and disposable . . .' (p. 24.)

The coordination (*Zuordnung*) of the ideational with the empirical world enables us to 'project the anticipated regularities of the practical Lebenswelt':

'Once one possesses the formulas, one possesses the *foresight* which is desired in practice'

– the foresight of that which is to be expected in the experience of concrete life (p. 43).

Husserl emphasizes the pre-scientific, technical connotations of mathematical exactness and fungibility. These central notions of modern science emerge, not as mere by-products of a pure science, but as pertaining to its inner conceptual structure. The scientific abstraction from concreteness, the quantification of qualities which yield exactness as well as universal validity, involve a specific concrete experience of the *Lebenswelt* – a specific mode of 'seeing' the world. And this 'seeing', in spite of its 'pure', disinterested character, is seeing within a purposive, practical context. It is anticipating (*Voraussehen*) and projecting (*Vorhaben*). Galilean science is the science of methodical, systematic anticipation and projection. But – and this is decisive – of a specific anticipation and projection – namely, that which experiences, comprehends, and shapes the world in terms of calculable, predictable relationships among exactly identifiable units. In this project, universal quantifiability is a prerequisite for the *domination* of nature. Individual, non-quantifiable qualities stand in the way of an organization of men and things in accordance with the measurable power to be extracted from them. But this is a specific, socio-historical project, and the consciousness which undertakes this project is the hidden subject of Galilean science; the latter is the technic, the art of anticipation extended in infinity (*ins Unendliche erweiterte Voraussicht:* p. 51).

Now precisely because Galilean science is, in the formation of its concepts, the technic of a specific *Lebenswelt*, it does not and cannot *transcend* this Lebenswelt. It remains essentially within the basic experiential framework and within the universe of ends set by this reality. In Husserl's formulation; in Galilean science, the 'concrete universe of causality becomes applied mathematics' (p. 122) – but the world of perception and experience, 'in which we live our whole practical life, remains as that which it is, in its essential structure, in its own concrete causality *unchanged* . . .' (p. 51; my italics).

A provocative statement, which is easily minimized, and I take the liberty of a possible overinterpretation. The statement does not refer simply to the fact that, in spite of non-Euclidean geometry, we still perceive and act in three-dimensional space; or that, in spite of the 'statistical' concept of causality, we still act, in common sense, in accord with the 'old' laws of causality. Nor does the statement contradict the perpetual changes in the world of daily practice as the result of 'applied mathematics'. Much more may be at stake: namely, the inherent limit of the established science and scientific method, by virtue of which they extend, rationalize, and insure the prevailing *Lebenswelt* without altering its existential structure – that is *without envisaging a qualitatively new mode of 'seeing'* and qualitatively new relations between men and between man and nature.

With respect to the institutionalized forms of life, science (pure as well as applied) would thus have a stabilizing, static, conservative function. Even its most revolutionary achievements would only be construction and destruction in line with a specific experience and organization of reality. The continuous self-correction of science – the revolution of its hypotheses which is built into its method – itself propels and extends the same historical universe, the same basic experience. It retains the same formal *a priori*, which makes for a very material, practical content. Far from minimizing the fundamental change which occurred with the establishment of Galilean science, Husserl's interpretation points up the radical break with the pre-Galilean tradition; the instrumentalist horizon of thought was indeed a new horizon. It created a new world of theoretical and practical Reason, but it has remained committed to a specific historical world which has its evident limits – in theory as well as in practice, in its pure as well as applied methods.

The preceding discussion seems to suggest not only the inner limitations and prejudices of scientific method but also its historical subjectivity. Moreover, it seems to imply the need for some sort of 'qualitative physics', revival of teleological philosophies, etc. I admit that this suspicion is justified, but at this point, I can only assert that no such obscurantist ideas are intended.

No matter how one defines truth and objectivity, they remain related to the human agents of theory and practice, and to their ability to comprehend and change their world. This ability in turn depends on the extent to which matter (whatever it may be) is recognized and understood as that which it *is* itself in all particular forms. In these terms, contemporary science is of immensely greater objective validity than its predecessors. One might even add that, at present, the scientific method is the only method that can claim such validity; the interplay of hypotheses and observable facts validates the hypotheses and establishes the facts. The point which I am trying to make is that science, *by virtue of its own method* and concepts, has projected and promoted a universe in which the domination of nature has remained linked to the domination of man – a link which tends to be fatal to this universe as a whole. Nature, scientifically comprehended and mastered, reappears in the technical apparatus of production and destruction which sustains and improves the life of the individuals while subordinating them to the masters of the apparatus. Thus the rational hierarchy merges with the social one. If this is the case, then the change in the direction of progress, which might sever this fatal link, would also affect the very structure of science – the scientific project. Its hypotheses, without losing their rational character, would develop in an essentially different experimental context (that of a pacified world); consequently, science would arrive at essentially different concepts of nature and establish essentially different facts. The rational society subverts the idea of Reason.

I have pointed out that the elements of this subversion, the notions of another rationality, were present in the history of thought from its beginning. The ancient idea of a state where Being attains fulfilment, where the tension between 'is' and 'ought' is resolved in the cycle of an eternal return, partakes of the metaphysics of domination. But it also pertains to the metaphysics of liberation – to the reconciliation of Logos and Eros. This idea envisages the coming-to-rest of the repressive productivity of Reason, the end of domination in gratification.

The two contrasting rationalities cannot simply be correlated with classical and modern thought respectively, as in John Dewey's formulation 'from contemplative enjoyment to active manipulation and control'; and 'from knowing as an esthetic enjoyment of the properties of nature . . . to knowing as a means of secular control'. (Dewey, 1929, pp. 95, 100.) Classical thought was sufficiently committed to the logic of secular control, and there is a sufficient component of indictment and refusal in modern thought to vitiate John Dewey's formulation. Reason, as conceptual thought and behaviour, is necessarily mastery, domination. Logos is law, rule, order, by virtue of knowledge. In subsuming particular cases under a universal, in subjecting it to their universal, thought attains mastery over the particular cases. It becomes capable not only of comprehending but also of acting upon them, controlling them. However, while all thought stands under the rule of logic, the unfolding of this logic is different in the various modes of thought. Classical formal and modern symbolic logic, transcendental and dialectical logic – each rules over a different universe of discourse and experience. They all developed within the historical continuum of domination to which they pay tribute. And this continuum bestows upon the modes of positive thinking their conformist and ideological character; upon those of negative thinking their speculative and utopian character.

By way of summary, we may now try to identify more clearly the hidden subject of scientific rationality and the hidden ends in its pure form. The scientific concept of a universally controllable nature projected nature as endless matter-in-function, the mere stuff of theory and practice. In this form, the object-world entered the construction of a technological universe – a universe of mental and physical instrumentalities, means in themselves. Thus it is a truly 'hypothetical' system, depending on a validating and verifying subject.

The processes of validation and verification may be purely theoretical ones, but they never occur in a vacuum and they never terminate in a private, individual mind. The hypothetical system of forms and functions becomes dependent on another system – a pre-established universe of ends, in which and *for* which it develops. What appeared extraneous, foreign to the theoretical project, shows forth as part of its very structure (method and concepts); pure objectivity reveals itself as *object for a subjectivity* which provides the Telos, the ends. In the construction of the technological reality, there is no such thing as a purely rational scientific order; the process of technological rationality is a political process.

Only in the medium of technology, man and nature become fungible objects of organization. The universal effectiveness and productivity of the apparatus under which they are subsumed veil the particular interests that organize the apparatus. In other words, technology has become the great vehicle of *reification* – reification in its most mature and effective form. The social position of the individual and his relation to others appear not only to be determined by objective qualities and laws, but these qualities and laws seem to lose their mysterious and uncontrollable character; they appear as calculable manifestations of (scientific) rationality. The world tends to become the stuff of total administration, which absorbs even the administrators. The web of domination has become the web of Reason itself, and this society is fatally entangled in it. And the transcending modes of thought seem to transcend Reason itself.

Under these conditions, scientific thought (scientific in the larger sense, as opposed to muddled, metaphysical, emotional, illogical thinking) outside the physical science assumes the form of a pure and self-contained formalism (symbolism) on the one hand, and a total empiricism on the other. (The contrast is not a conflict. See the very empirical application of mathematics and symbolic logic in electronic industries.) In relation to the established universe of discourse and behaviour, non-contradiction and non-transcendance is the common denominator. Total empiricism reveals its ideological function in contemporary philosophy. With respect to this function, some aspects of linguistic analysis will be discussed in the following chapter. This discussion is to prepare the ground for the attempt to show the barriers which prevent this empiricism from coming to grips with reality, and establishing (or rather re-establishing) the concepts which may break these barriers.

References

BACHELARD, G. (1951), *L'Activité Rationaliste de la Physique Contemporaire*, Presses Universitaires, Paris.

BACHELARD, G. (1958), *La Conscience de Rationalité*, Presses Universitaires, Paris.
BIEMEL, W. (ed.) (1954), *Die Krisis der Europäischen Wissenschaften und die Tranzendentale Phänomenologie*, Haag, Nijhoff.
BUNGE, M. (1959), *Metascientific Queries*, Charles C. Thomas, Springfield, Ill.
DEWEY, J. (1929), *The Quest for Certainty*, New York.
DINGLER, H. (1951), 'Philosophy of physics, 1850–1950', *Nature*, vol. 168, p. 636.
GRÜNBAUM, A. (1954), 'Operationalism and relativity', in P. G. Frank (ed.), *The Validation of Scientific Theories*, Becon Press, pp. 84–94.
HEIDEGGER, M. (1950), *Holzwege*, Klostermann, Frankfurt.
HEIDEGGER, M. (1954), *Vortiöge and Aufsätze*, Gunther Neske, Pfüllingen.
HEISENBERG, W. (1948), 'Uber den Bengriff "Abgeschlossene Theorie"', *Dialectica*, vol. 2, no. 1, p. 333.
HEISENBERG, W. (1958), *The Physicist's Concept of Nature*, Hutchinson.
HEISENBERG, W (1959), *Physics and Philosophy*, Allen & Unwin.
HORKHEIMER, M., and ADORNO, T. W. (1947), *Dialektik der Aufklarung*, Amsterdam.
MARX, K. (1935), 'The poverty of philosophy', ch. 2, 'Second Observation', in E. Burns (ed.), *A Handbook of Marxism*, New York.
MARX, K., and ENGELS, F. (1846), *Die Deutsche Ideologie*, trans. Molitor,
PIAGET, J. (1950), *Introduction a l'Epistémologie Génétique*, vol. 3, Presses Universitaires, Paris.
POPPER, K. (1957), *British Philosophy in the Mid-Century*, C. A. Mace (ed.), Macmillan Co.
QUINE, W. V. O. (1953), *From a Logical Point of View*, Harvard University Press.
SIMONDON, G. (1958), *Du Mode d'Existence des Objects Techniques*, Autier, Paris.
WEIZSACKER, C. F. Von (1949), *The History of Nature*, Chicago University Press.

20 Jürgen Habermas

Science and Technology as Ideology

Excerpt from Jürgen Habermas, *Toward a Rational Society*, translated by J. J. Shapiro, Heinemann, 1971. First published in German in 1968.

By means of the concept of 'rationalization' Weber attempted to grasp the repercussions of scientific-technical progress on the institutional framework of societies engaged in 'modernization'. He shared this interest with the classical sociological tradition in general, whose pairs of polar concepts all revolve about the same problem: how to construct a conceptual model of the institutional change brought about by the extension of subsystems of purposive-rational action. Status and contract, *Gemeinschaft* and *Gesellschaft*, mechanical and organic solidarity, informal and formal groups, primary and secondary groups, culture and civilization, traditional and bureaucratic authority, sacral and secular associations, military and industrial society, status group and class – all of these pairs of concepts represent as many attempts to grasp the structural change of the institutional framework of a traditional society on the way to becoming a modern one. Even Parsons' catalog of possible alternatives of value-orientations belongs in the list of these attempts, although he would not admit it. Parsons claims that his list systematically represents the decisions between alternative value-orientations that must be made by the subject of any action whatsoever, regardless of the particular or historical context. But if one examines the list, one can scarcely overlook the historical situation of the inquiry on which it is based. The four pairs of alternative value-orientations,

affectivity versus *affective neutrality*
particularism versus *universalism*
ascription versus *achievement*
diffuseness versus *specificity*

which are supposed to take into account *all* possible fundamental decisions, are tailored to an analysis of *one* historical process. In fact they define the relative dimensions of the modification of dominant attitudes in the transition from traditional to modern society. Subsystems of purposive-rational action do indeed demand orientation to the postponement of gratification, universal norms, individual achievement and active mastery, and specific and analytic relationships, rather than to the opposite orientations.

In order to reformulate what Weber called 'rationalization', I should like to go beyond the subjective approach that Parsons shares with Weber and propose another categorial framework. I shall take as my starting point the fundamental distinction between *work* and *interaction*.[1]

By 'work' or *purposive-rational action* I understand either instrumental action or rational choice or their conjunction. Instrumental action is governed by *technical rules* based on empirical knowledge. In every case they imply conditional predictions about observable events, physical or social. These predictions can prove correct or incorrect. The conduct of rational choice is governed by *strategies* based on analytic knowledge. They imply deductions from preference rules (value systems) and decision procedures; these propositions are either correctly or incorrectly deduced. Purposive-rational action realizes defined goals under given conditions. But while instrumental action organizes means that are appropriate or inappropriate according to criteria of an effective control of reality, strategic action depends only on the correct evaluation of possible alternative choices, which results from calculation supplemented by values and maxims.

By 'interaction', on the other hand, I understand *communicative action*, symbolic interaction. It is governed by binding *consensual norms*, which define reciprocal expectations about behavior and which must be understood and recognized by at least two acting subjects. Social norms are enforced through sanctions. Their meaning is objectified in ordinary language communication. While the validity of technical rules and strategies depends on that

1. On the context of these concepts in the history of philosophy see my contribution to the *Festschrift* for Karl Löwith. (Habermas, 1967a.)

of empirically true or analytically correct propositions, the validity of social norms is grounded only in the intersubjectivity of the mutual understanding of intentions and secured by the general recognition of obligations. Violation of a rule has a different consequence according to type. *Incompetent* behavior, which violates valid technical rules or strategies, is condemned *per se* to failure through lack of success; the 'punishment' is built, so to speak, into its rebuff by reality. *Deviant* behavior, which violates consensual norms, provokes sanctions that are connected with the rules only externally, that is by convention. Learned rules of purposive-rational action supply us with *skills*, internalized norms with *personality structures*. Skills put us in a position to solve problems; motivations allow us to follow norms. The diagram below

	Institutional framework: symbolic interaction	Systems of purposive-rational (instrumental and strategic) action
action-orienting rules	social norms	technical rules
level of definition	intersubjectively shared ordinary language	context-free language
type of definition	reciprocal expectations about behavior	conditional predictions conditional imperatives
mechanisms of acquisition	role internalization	learning of skills and qualifications
function of action type	maintenance of institutions (conformity to norms on the basis of reciprocal enforcement)	problem-solving (goal attainment, defined in means-ends relations)
sanctions against violation of rules	punishment on the basis of conventional sanctions: failure against authority	inefficacy: failure in reality
'rationalization'	emancipation, individuation; extension of communication free of domination	growth of productive forces; extension of power of technical control

summarizes these definitions. They demand a more precise explanation, which I cannot give here. It is above all the bottom column which I am neglecting here, and it refers to the very problem for whose solution I am introducing the distinction between work and interaction.

In terms of the two types of action we can distinguish between social systems according to whether purposive-rational action or interaction predominates. The institutional framework of a society consists of norms that guide symbolic interaction. But there are subsystems such as (to keep to Weber's examples) the economic system or the state apparatus, in which primarily sets of purposive-rational action are institutionalized. These contrast with subsystems such as family and kinship structures, which, although linked to a number of tasks and skills, are primarily based on moral rules of interaction. So I shall distinguish generally at the analytic level between the *institutional framework* of a society or the sociocultural life-world and the *subsystems of purposive-rational action* that are 'embedded' in it. Insofar as actions are determined by the institutional framework they are both guided and enforced by norms. Insofar as they are determined by subsystems of purposive-rational action, they conform to patterns of instrumental or strategic action. Of course, only institutionalization can guarantee that such action will in fact follow definite technical rules and expected strategies with adequate probability.

With the help of these distinctions we can reformulate Weber's concept of 'rationalization'.

The term 'traditional society' has come to denote all social systems that generally meet the criteria of civilizations. The latter represent a specific stage in the evolution of the human species. They differ in several traits from more primitive social forms: a centralized ruling power (state organization of political power in contrast to tribal organization); the division of society into socioeconomic classes (distribution to individuals of social obligations and rewards according to class membership and not according to kinship status); the prevalence of a central worldview (myth, complex religion) to the end of legitimating political power (thus converting power into authority). Civilizations are established on the basis of a relatively developed technology and of division of labor in the social process of production, which make possible a surplus product, i.e. a quantity of goods exceeding that needed for the satisfaction of immediate and elementary needs. They owe their existence to the solution of the problem that first arises with the production of a surplus product, namely, how to distribute wealth and labor both unequally and yet legitimately according to criteria other than those generated by a kinship system. (Lenski, 1966.)

In our context it is relevant that despite considerable differences in their level of development, civilizations, based on an economy dependent on agriculture and craft production, have tolerated technical innovation and organizational improvement only within definite limits. One indicator of the traditional limits to the development of the forces of production is that until about three hundred years ago no major social system had produced more than the equivalent of a maximum of two hundred dollars per capita per annum. The stable pattern of a precapitalist mode of production, preindustrial technology, and premodern science makes possible a typical relation of the institutional framework to subsystems of purposive-rational action. For despite considerable progress, these subsystems, developing out of the system of social labor and its stock of accumulated technically exploitable knowledge, never reached that measure of extension after which their 'rationality' would have become an open threat to the authority of the cultural traditions that legitimate political power. The expression 'traditional society' refers to the circumstance that the institutional framework is grounded in the unquestionable underpinning of legitimation constituted by mythical, religious or metaphysical interpretations of reality – cosmic as well as social – as a whole. 'Traditional' societies exist as long as the development of subsystems of purposive-rational action keep within the limits of the legitimating efficacy of cultural traditions. (See Berger, 1967.) This is the basis for the 'superiority' of the institutional framework, which does not preclude structural changes adapted to a potential surplus generated in the economic system but does preclude critically challenging the traditional form of legitimation. This immunity is a meaningful criterion for the delimitation of traditional societies from those which have crossed the threshold to modernization.

The 'superiority criterion', consequently, is applicable to all forms of class society organized as a state in which principles of universally valid rationality (whether of technical or strategic means-ends relations) have not explicitly and successfully called into question the cultural validity of intersubjectively shared traditions, which function as legitimations of the political system. It is only since the capitalist mode of production has equipped the economic system with a self-propelling mechanism that ensures long-term continuous growth (despite crises) in the productivity of labor that the introduction of new technologies and strategies, i.e. innovation as such, has been institutionalized. As Marx and Schumpeter have proposed in their respective theories, the capitalist mode of production can be comprehended as a mechanism that guarantees the *permanent* expansion of subsystems of purposive-rational action and thereby overturns the traditionalist 'superiority' of the institutional framework to the forces of production. Capitalism is the first mode of production in world history to institutionalize self-sustaining economic growth. It has generated an industrial system that could be freed from the institutional framework of capitalism and connected to mechanisms other than that of the utilization of capital in private form.

What characterizes the passage from traditional society to society commencing the process of modernization is *not* that structural modification of the institutional framework is necessitated under the pressure of relatively developed productive forces, for that is the mechanism of the evolution of the species from the very beginning. What is new is a level of development of the productive forces that makes permanent the extension of subsystems of purposive-rational action and thereby calls into question the traditional form of the legitimation of power. The older mythic, religious, and metaphysical worldviews obey the logic of interaction contexts. They answer the central questions of men's collective existence and of individual life history. Their themes are justice and freedom, violence and oppression, happiness and gratification, poverty, illness, and death. Their categories are victory and defeat, love and hate, salvation and damnation. Their logic accords with the grammar of systematically distorted communication and with the fateful causality of dissociated symbols and suppressed motives. (See Habermas, 1968.) The rationality of language games, associated with communicative action, is confronted at the threshold of the modern period with the rationality of means-ends relations, associated with instrumental and strategic action. As soon as this confrontation can arise, the end of traditional society is in sight: the traditional form of legitimation breaks down.

Capitalism is defined by a mode of production that not only poses this problem but also solves it. It provides a legitimation of domination which is no longer called down from the lofty heights of cultural tradition but instead summoned up from the base of social labor. The institution of the market, in which private property owners exchange commodities – including the market on which propertyless private individuals exchange their labor power as their only commodity – promises that exchange relations will be and are just owing to equivalence. Even this bourgeois ideology of justice, by adopting the category of reciprocity, still employs a relation of communicative action as the basis of legitimation. But the principle of reciprocity is now the organizing principle of the sphere of production and reproduction itself. Thus on the base of a market economy, political domination can be legitimated henceforth 'from below' rather than 'from above' (through invocation of cultural tradition).

If we suppose that the division of society into socioeconomic classes derives from the differential distribution among social groups of the relevant means of production, and that this distribution itself is based on the institutionalization of relations of social force, then we may assume that in all civilizations this institutional framework has been identical with the system of political domination: traditional authority was political authority. Only with the emergence of the capitalist mode of production can the legitimation of the institutional framework be linked immediately with the system of social labor. Only then can the property order change from a *political relation* to a *production relation*, because it legitimates itself through the rationality of the market, the ideology of exchange society, and no longer through a legitimate power structure. It is now the political system which is justified in terms of the legitimate relations of production: this is the real meaning and function of rationalist natural law from Locke to Kant. (See also, Strauss, 1963; MacPherson, 1962; Habermas, 1967.) The institutional framework of society is only mediately political and immediately economic (the bourgeois constitutional state as 'superstructure').

The superiority of the capitalist mode of production to its predecessors has these two roots: the establishment of an economic mechanism that renders permanent the expansion of subsystems of purposive-rational action, and the creation of an economic legitimation by means of which the political system can be adapted to the new requisites of rationality brought about by these developing subsystems. It is this process of adaptation that Weber comprehends as 'rationalization'. Within it we can distinguish between two tendencies: rationalization 'from below' and rationalization 'from above'.

A permanent pressure for adaptation arises from below as soon as the new mode of production becomes fully operative through the institutionalization of a domestic market for goods and labor power and of the capitalist enterprise. In the system of social labor this institutionalization ensures cumulative progress in the forces of production and an ensuing horizontal extension of subsystems of purposive-rational action – at the cost of economic crises, to be sure. In this way traditional structures are increasingly subordinated to conditions of instrumental or strategic rationality: the organization of labor and of trade, the network of transportation, information, and communication, the institutions of private law, and, starting with financial administration, the state bureaucracy. Thus arises the substructure of a society under the compulsion of modernization. The latter eventually widens to take in all areas of life: the army, the school system, health services, and even the family. Whether in city or country, it induces an urbanization of the *form* of life. That is, it generates subcultures that train the individual to be able to 'switch over' at any moment from an interaction context to purposive-rational action.

This pressure for rationalization coming from below is met by a compulsion to rationalize coming from above. For, measured against the new standards of purposive rationality, the power-legitimating and action-orienting traditions – especially mythological interpretations and religious world views – lost their cogency. On this level of generalization, what Weber termed 'secularization' has two aspects. First, traditional worldviews and objectivations lose their power and validity *as* myth, *as* public religion, *as* customary ritual, *as* justifying metaphysics, *as* unquestionable tradition. Instead, they are reshaped into subjective belief systems and ethics which ensure the private cogency of modern value-orientations (the 'Protestant ethic'). Second, they are transformed into constructions that do both at once: criticize tradition and reorganize the released material of tradition according to the principles of formal law and the exchange of equivalents (rationalist natural law). Having become fragile, existing legitimations are replaced by new ones. The latter emerge from the critique of the dogmatism of traditional interpretations of the world and claim a scientific character. Yet they retain legitimating functions, thereby keeping actual power relations inaccessible to analysis and to public consciousness. It is in this way that ideologies in the restricted sense first came into being. They replace traditional legitimations of power by appearing in the mantle of modern science and by deriving their justification from the critique of ideology. Ideologies are coeval with the critique of ideology. In this sense there can be no prebourgeois 'ideologies'.

In this connection modern science assumes a singular function. In distinction from the philosophical sciences of the older sort, the empirical sciences have developed since Galileo's time within a methodological frame of reference that reflects the transcendental viewpoint of possible technical control. Hence the modern sciences produce knowledge which through its *form* (and not through the subjective intention of scientists) is technically exploitable knowledge, although the possible applications generally are realized afterwards. Science and technology were not interdependent until late into the nineteenth century. Until then modern science did not contribute to the acceleration of technical development nor, consequently, to the pressure toward rationalization from below. Rather, its contribution to the modernization process was indirect. Modern physics gave rise to a philosophical approach that interpreted nature and society according to a model borrowed from the natural sciences and induced, so to speak, the mechanistic worldview of the seventeenth century. The reconstruction of classical natural law was carried out in this framework. This modern natural law was the basis of the bourgeois revolutions of the seventeenth, eighteenth, and nineteenth centuries, through which the old legitimations of the power structure were finally destroyed. (See Habermas, 1967.)

By the middle of the nineteenth century the capitalist mode of production had developed so fully in England and France that Marx was able to identify the locus of the institutional framework of society in the relations of production and at the same time criticize the legitimating basis constituted by the exchange of equivalents. He carried out the critique of bourgeois ideology in the form of *political economy*. His labor theory of value destroyed the semblance of freedom, by means of which the legal institution of the free labor contract had made unrecognizable the relationship of social force that underlay the wage-labor relationship. Marcuse's criticism of Weber is that the latter, disregarding this Marxian insight, upholds an abstract concept of rationalization, which not merely fails to express the specific class content of the adaptation of the institutional framework to the developing systems of purposive-rational action, but conceals it. Marcuse knows that the Marxian analysis can no longer be applied as it stands to advanced capitalist society, with which Weber was already confronted. But he wants to show through the example of Weber that the evolution of modern society in the framework of state-regulated capitalism cannot be conceptualized if liberal capitalism has not been analyzed adequately.

Since the last quarter of the nineteenth century two developmental tendencies have become noticeable in the most advanced capitalist countries: an increase in state intervention in order to secure the system's stability, and a growing interdependence of research and technology, which has turned the sciences into the leading productive force. Both tendencies have destroyed the particular constellation of institutional framework and subsystems of purposive-rational action which characterized liberal capitalism, thereby eliminating the conditions relevant for the application of political economy in the version correctly formulated by Marx for liberal capitalism. I believe that Marcuse's basic thesis, according to which technology and science today also take on the function of legitimating political power, is the key to analyzing the changed constellation.

The permanent regulation of the economic process by means of state intervention arose as a defense mechanism against the dysfunctional tendencies, which threaten the system, that capitalism generates when left to itself. Capitalism's actual development manifestly contradicted the capitalist idea of a bourgeois society, emancipated from domination, in which power is neutralized. The root ideology of just exchange, which Marx unmasked in theory, collapsed in practice. The form of capital utilization through private ownership could only be maintained by the governmental corrective of a social and economic policy that stabilized the business cycle. The institutional framework of society was repoliticized. It no longer coincides immediately with the relations of production, i.e. with an order of private law that secures capitalist economic activity and the corresponding general guarantees of order provided by the bourgeois state. But this means a change in the relation of the economy to the political system: politics is no longer *only* a phenomenon of the superstructure. If society no longer 'autonomously' perpetuates itself through self-regulation as a sphere preceding and lying at the basis of the state – and its ability to do so was the really novel feature of the capitalist mode of production – then society and the state are no longer in the relationship that Marxian theory had defined as that of base and superstructure. Then, however, a critical theory of society can no longer be constructed in the exclusive form of a critique of political economy. A point of view that methodically isolates the economic laws of motion of society can claim to grasp the overall structure of social life in its essential categories only as long as politics depends on the economic base. It becomes inapplicable when the 'base' has to be comprehended as in itself a function of governmental activity and political conflicts. According to Marx, the critique of political economy was the theory of bourgeois society only as *critique of ideology*. If, however, the ideology of just exchange disintegrates, then the power structure can no longer be criticized *immediately* at the level of the relations of production.

With the collapse of this ideology, political power requires a new legitimation. Now since the power indirectly exercised over the exchange process is itself operating under political control and state regulation, legitimation can no longer be derived from the unpolitical order constituted by the relations of production. To this extent the requirement for direct legitimation, which exists in precapitalist societies, reappears. On the other hand, the resuscitation of immediate political domination (in the traditional form of legitimation on the basis of cosmological worldviews) has become impossible. For traditions have already been disempowered. Moreoever, in industrially developed societies the results of bourgeois emancipation from immediate political domination (civil and political rights and the mechanism of general elections) can be fully ignored only in periods of reaction. Formally democratic government in systems of state-regulated capitalism is subject to a need for legitimation which cannot be met by a return to a prebourgeois form. Hence the ideology of free exchange is replaced by a substitute program. The latter is oriented not to the social results of the institution of the market but to those of government action designed to compensate for the dysfunctions of free exchange. This policy combines the element of the bourgeois ideology of achievement (which, however, displaces assignment of status according to the standard of individual achievement from the market to the school system) with a guaranteed minimum level of welfare, which offers secure employment and a stable income. This substitute program obliges the political system to maintain stabilizing conditions for an economy that guards against risks to growth and guarantees social security and the chance for individual upward mobility. What is needed to this end is latitude for manipulation by state interventions that, at the cost of limiting the institutions of private law, secure the private form of capital utilization *and bind the masses' loyalty to this form.*

Insofar as government action is directed toward the economic system's stability and growth, politics now takes on a peculiarly negative character. For it is oriented toward the elimination of dysfunctions and the avoidance of risks that threaten the system: not, in other words, toward the *realization of practical goals* but toward the *solution of technical problems.* Claus Offe pointed this out in his paper at the 1968 Frankfurt Sociological Conference:

In this structure of the relation of economy and the state, 'politics' degenerates into action that follows numerous and continually emerging 'avoidance imperatives': the mass of differentiated social-scientific information that flows into the political system allows both the early identification of risk zones and the treatment of actual dangers. What is new about this structure is . . . that the risks to stability built into the mechanism of private capital utilization in highly organized markets, risks that can be manipulated, prescribe preventive actions and measures that *must* be accepted as long as they are to accord with the existing legitimation resources (i.e., substitute program).[2]

Offe perceives that through these preventive action-orientations, government activity is restricted to administratively soluble technical problems, so that practical questions evaporate, so to speak. *Practical substance is eliminated.*[3]

Old-style politics was forced, merely through its traditional form of legitimation, to define itself in relation to practical goals: the 'good life' was interpreted in a context defined by interaction relations. The same still held for the ideology of bourgeois society. The substitute program prevailing today, in contrast, is aimed exclusively at the functioning of a manipulated system. It eliminates practical questions and therewith precludes discussion about the adoption of standards; the latter could emerge only from a democratic decision-making process. The solution of technical problems is not dependent on public discussion. Rather, public discussions could render problematic the framework within which the tasks of government action present themselves as technical ones. Therefore the new politics of state interventionism requires a depoliticization of the mass of the population. To the extent that practical questions are eliminated, the public realm also loses its political function. At the same time, the institutional framework of society is still distinct from the systems of purposive-rational action themselves. Its organization continues to be a problem of *practice* linked to communication, not one of *technology*, no matter how scientifically guided. Hence the bracketing, out of practice associated with the new kind of politics is not automatic. The substitute program, which legitimates power today, leaves unfilled a vital need for legitimation: how will the depoliticization of the masses be made plausible to them? Marcuse would be able to answer: by having technology and science *also* take on the role of an ideology.

2. Offe (1969). The quotation in the text is from the original manuscript, which differs in formulation from the published text.

3. See Terminological note on page 374.

Since the end of the ninteenth century the other developmental tendency characteristic of advanced capitalism has become increasingly momentous: the scientization of technology. The institutional pressure to augment the productivity of labor through the introduction of new technology has always existed under capitalism. But innovations depended on sporadic inventions, which, while economically motivated, were still fortuitous in character. This changed as technical development entered into a feedback relation with the progress of the modern sciences. With the advent of large-scale industrial research, science, technology, and industrial utilization were fused into a system. Since then, industrial research has been linked up with research under government contract, which primarily promotes scientific and technical progress in the military sector. From there information flows back into the sectors of civilian production. Thus technology and science become a leading productive force, rendering inoperative the conditions for Marx's labor theory of value. It is no longer meaningful to calculate the amount of capital investment in research and development on the basis of the value of unskilled (simple) labor power, when scientific-technical progress has become an independent source of surplus value, in relation to which the only source of surplus value considered by Marx, namely the labor power of the immediate producers, plays an ever smaller role.[4]

As long as the productive forces were visibly linked to the rational decisions and instrumental action of men engaged in social production, they could be understood as the potential for a growing power of technical control and not be confused with the institutional framework in which they are embedded. However, with the institutionalization of scientific-technical progress, the potential of the productive forces has assumed a form owing to which men lose consciousness of the dualism of work and interaction.

It is true that social interests still determine the direction, functions, and pace of technical progress. But these interests define the social system so much as a whole that they coincide with the interest in maintaining the system. *As such* the private form of capital utilization and a distribution mechanism for social rewards

4. The most recent explication of this is Löbl (1968).

that guarantees the loyalty of the masses are removed from discussion. The quasi-autonomous progress of science and technology then appears as an independent variable on which the most important single system variable, namely economic growth, depends. Thus arises a perspective in which the development of the social system *seems* to be determined by the logic of scientific-technical progress. The immanent law of this progress seems to produce objective exigencies, which must be obeyed by any politics oriented toward functional needs. But when this semblance has taken root effectively, then propaganda can refer to the role of technology and science in order to explain and legitimate why in modern societies the process of democratic decision-making about practical problems loses its function and 'must' be replaced by plebiscitary decisions about alternative sets of leaders of administrative personnel. This technocracy thesis has been worked out in several versions on the intellectual level. (See Schelsky, 1961; Ellul, 1967; Gehlen, 1963, 1964.) What seems to me more important is that it can also become a background ideology that penetrates into the consciousness of the depoliticized mass of the population, where it can take on legitimating power.[5] It is a singular achievement of this ideology to detach society's self-understanding from the frame of reference of communicative action and from the concepts of symbolic interaction and replace it with a scientific model. Accordingly the culturally defined self-understanding of a social life-world is replaced by the self-reification of men under categories of purposive-rational action and adaptive behavior.

The model according to which the planned reconstruction of society is to proceed is taken from systems analysis. It is possible in principle to comprehend and analyze individual enterprises and organizations, even political or economic subsystems and social systems as a whole, according to the pattern of self-regulated systems. It makes a difference, of course, whether we use a cybernetic frame of reference for analytic purposes or *organize* a given social system in accordance with this pattern as a man-machine system. But the transferral of the analytic model to the level of

5. To my knowledge there are no empirical studies concerned specifically with the propagation of this background ideology. We are dependent on extrapolations from the findings of other investigations.

social organization is implied by the very approach taken by systems analysis. Carrying out this intention of an instinct-like self-stabilization of social systems yields the peculiar perspective that the structure of one of the two types of action, namely the behavioral system of purposive-rational action, not only predominates over the institutional framework but gradually absorbs communicative action as such. If, with Arnold Gehlen, one were to see the inner logic of technical development as the step-by-step disconnection of the behavioral system of purposive-rational action from the human organism and its transferral to machines, then the technocratic intention could be understood as the last stage of this development. For the first time man can not only, as *homo faber*, completely objectify himself and confront the achievements that have taken on independent life in his products; he can in addition, as *homo fabricatus*, be integrated into his technical apparatus if the structure of purposive-rational action can be successfully reproduced on the level of social systems. According to this idea the institutional framework of society – which previously was rooted in a different type of action – would now, in a fundamental reversal, be *absorbed* by the subsystems of purposive-rational action, which were embedded in it.

Of course this technocratic intention has not been realized anywhere even in its beginnings. But it serves as an ideology for the new politics, which is adapted to technical problems and brackets out practical questions. Furthermore it does correspond to certain developmental tendencies that could lead to a creeping erosion of what we have called the institutional framework. The manifest domination of the authoritarian state gives way to the manipulative compulsions of technical-operational administration. The moral realization of a normative order is a function of communicative action oriented to shared cultural meaning and presupposing the internalization of values. It is increasingly supplanted by conditioned behavior, while large organizations as such are increasingly patterned after the structure of purposive-rational action. The industrially most advanced societies seem to approximate the model of behavioral control steered by external stimuli rather than guided by norms. Indirect control through fabricated stimuli has increased, especially in areas of putative subjective freedom (such as electoral, consumer, and leisure behavior). Sociopsychologically, the era is typified less by the authoritarian personality than by the destructuring of the superego. The increase in *adaptive behavior* is, however, only the obverse of the dissolution of the sphere of linguistically mediated interaction by the structure of purposive-rational action. This is paralleled subjectively by the disappearance of the difference between purposive-rational action and interaction from the consciousness not only of the sciences of man, but of men themselves. The concealment of this difference proves the ideological power of the technocratic consciousness.

In consequence of the two tendencies that have been discussed, capitalist society has changed to the point where two key categories of Marxian theory, namely class struggle and ideology, can no longer be employed as they stand.

It was on the basis of the capitalist mode of production that the struggle of social classes as such was first constituted, thereby creating an objective situation from which the class structure of traditional society, with its immediately political constitution, could be *recognized* in retrospect. State-regulated capitalism, which emerged from a reaction against the dangers to the system produced by open class antagonism, suspends class conflict. The system of advanced capitalism is so defined by a policy of securing the loyalty of the wage-earning masses through rewards, that is, by avoiding conflict, that the conflict still built into the structure of society in virtue of the private mode of capital utilization is the very area of conflict which has the greatest probability of remaining latent. It recedes behind others, which, while conditioned by the mode of production, can no longer assume the form of class conflicts. In the paper cited, Claus Offe has analyzed this paradoxical state of affairs, showing that open conflicts about social interests break out with greater probability the less their frustration has dangerous consequences for the system. The needs with the greatest conflict potential are those on the periphery of the area of state intervention. They are far from the central conflict being kept in a state of latency and therefore they are not seen as having priority among dangers to be warded off. Conflicts are set off by these needs to the extent that disproportionately scattered state interventions produce backward areas of development and corresponding disparity tensions:

The disparity between areas of life grows above all in view of the differential state of development obtaining between the actually institutionalized and the possible level of technical and social progress. The disproportion between the most modern apparatuses for industrial and military purposes and the stagnating organization of the transport, health, and educational systems is just as well known an example of this disparity between areas of life as is the contradiction between rational planning and regulation in taxation and finance policy and the unplanned, haphazard development of cities and regions. Such contradictions can no longer be designated accurately as antagonisms between classes, yet they can still be interpreted as results of the still dominant process of the private utilization of capital and of a specifically capitalist power structure. In this process the prevailing interests are those which, without being clearly localizable, are in a position, on the basis of the established mechanism of the capitalist economy, to react to disturbances of the conditions of their stability by producing risks relevant to the system as a whole. (1969).

The interests bearing on the maintenance of the mode of production can no longer be 'clearly localized' in the social system as class interests. For the power structure, aimed as it is at avoiding dangers to the system, precisely excludes 'domination' (as immediate political or economically mediated social force) exercised in such a manner that one class subject *confronts* another as an identifiable group.

This means not that class antagonisms have been abolished but that they have become *latent*. Class distinctions persist in the form of subcultural traditions and corresponding differences not only in the standard of living and life style but also in political attitude. The social structure also makes it probable that the class of wage earners will be hit harder than other groups by social disparities. And finally, the generalized interest in perpetuating the system is still anchored today, on the level of immediate life chances, in a structure of privilege. The concept of an interest that has become *completely* independent of living subjects would cancel itself out. But with the deflection of dangers to the system in state-regulated capitalism, the political system has incorporated an interest – which transcends latent class boundaries – in preserving the compensatory distribution façade.

Furthermore, the displacement of the conflict zone from the class boundary to the underprivileged regions of life does not mean at all that serious conflict potential has been disposed of. As the extreme example of racial conflict in the United States shows, so many consequences of disparity can accumulate in certain areas and groups that explosions resembling civil war can occur. But unless they are connected with protest potential from other sectors of society no conflicts arising from such underprivilege can really overturn the system – they can only provoke it to sharp reactions incompatible with formal democracy. For underprivileged groups are not social classes, nor do they ever even potentially represent the mass of the population. Their *disfranchisement* and pauperization no longer coincide with *exploitation*, because the system does not live off their labor. They can represent at most a past phase of exploitation. But they cannot through the withdrawal of cooperation attain the demands that they legitimately put forward. That is why these demands retain an appelative character. In the case of long-term nonconsideration of their legitimate demands underprivileged groups can in extreme situations react with desperate destruction and self-destruction. But as long as no coalitions are made with privileged groups, such a civil war lacks the chance of revolutionary success that class struggle possesses.

With a series of restrictions this model seems applicable even to the relations between the industrially advanced nations and the formerly colonial areas of the Third World. Here, too, growing disparity leads to a form of underprivilege that in the future surely will be increasingly less comprehensible through categories of exploitation. Economic interests are replaced on this level, however, with immediately military ones.

Be that as it may, in advanced capitalist society deprived and privileged groups no longer confront each other *as* socioeconomic classes – and to some extent the boundaries of underprivilege are no longer even specific to groups and instead run across population categories. Thus the fundamental relation that existed in all traditional societies and that came to the fore under liberal capitalism is mediatized, namely the class antagonism between partners who stand in an institutionalized relationship of force, economic exploitation, and political oppression to one another, and in which communication is so distorted and restricted that the legitimations serving as an ideological veil cannot be called into question. Hegel's concept of the ethical totality of a living relationship which is sundered because one subject does not reciprocally satisfy the needs of the other is no longer an appropriate model for the mediatized class structure of organized, advanced capitalism. The suspended dialectic of the ethical generates the peculiar semblance of *post-histoire*. The reason is that relative growth of the productive forces no longer represents *eo ipso* a potential that points beyond the existing framework with emancipatory consequences, in view of which legitimations of an existing power structure become enfeebled. For the leading productive force – controlled scientific-technical progress itself – has now become the basis of legitimation. Yet this new form of legitimation has cast off the old shape of *ideology*.

Technocratic consciousness is, on the one hand, 'less ideological' than all previous ideologies. For it does not have the opaque force of a delusion that only transfigures the implementation of interests. On the other hand today's dominant, rather glassy background ideology, which makes a fetish of science, is more irresistible and farther-reaching than ideologies of the old type. For with the veiling of practical problems it not only justifies a *particular class*'s interest in domination and represses *another class*'s partial need for emancipation, but affects the human race's emancipatory interest as such.

Technocratic consciousness is not a rationalized, wish-fulfilling fantasy, not an 'illusion' in Freud's sense, in which a system of interaction is either represented or interpreted and grounded. Even bourgeois ideologies could be traced back to a basic pattern of just interactions, free of domination and mutually satisfactory. It was these ideologies which met the criteria of wish-fulfillment and substitute gratification; the communication on which they were based was so limited by repressions that the relation of force once institutionalized as the capital-labor relation could not even be called by name. But the technocratic consciousness is not based in the same way on the causality of dissociated symbols and unconscious motives, which generates both false consciousness and the power of reflection to which the critique of ideology is indebted. It is less vulnerable to reflection, because it is no longer *only* ideology. For it does not, in the manner of ideology, express a projection of the 'good life' (which even if not identifiable with a bad reality, can at least be brought into virtually satisfactory accord with it). Of course the new ideology, like the old, serves to impede making the foundations of society the object of thought and reflection. Previously, social force lay at the basis of the relation between capitalist and wage-laborers. Today the basis is provided by structural conditions which predefine the tasks of system maintenance: the private form of capital utilization and a political form of distributing social rewards that guarantees mass loyalty. However, the old and new ideology differ in two ways.

First, the capital-labor relation today, because of its linkage to a loyalty-ensuring political distribution mechanism, no longer engenders uncorrected exploitation and oppression. The process through which the persisting class antagonism has been made virtual presupposes that the repression on which the latter is based first came to consciousness in history and *only then* was stabilized in a modified form as a property of the system. Technocratic consciousness, therefore, cannot rest in the same way on collective repression as did earlier ideologies. Second, mass loyalty today is created only with the aid of rewards for *privatized needs*. The achievements in virtue of which the system justifies itself may not in principle be interpreted politically. The acceptable interpretation is immediately in terms of allocations of money and leisure time (neutral with regard to their use), and mediately in terms of the technocratic justification of the occlusion of practical questions. Hence the new ideology is distinguished from its predecessor in that it severs the criteria for justifying the organization of social life from any normative regulation of interaction, thus depoliticizing them. It anchors them instead in functions of a putative system of purposive-rational action.

Technocratic consciousness reflects not the sundering of an ethical situation but the repression of 'ethics' as such as a category of life. The common, positivist way of thinking renders inert the frame of reference of interaction in ordinary language, in which domination and ideology both arise under conditions of distorted communication and can be reflectively detected and broken down.

The depoliticization of the mass of the population, which is legitimated through technocratic consciousness, is at the same time men's self-objectification in categories equally of both purposive-rational action and adaptive behavior. The reified models of the sciences migrate into sociocultural life-world and gain objective power over the latter's self-understanding. The ideological nucleus of this consciousness is *the elimination of the distinction between the practical and the technical.* It reflects, but does not objectively account for, the new constellation of a disempowered institutional framework and systems of purposive-rational action that have taken on a life of their own.

The new ideology consequently violates an interest grounded in one of the two fundamental conditions of our cultural existence: in language, or more precisely, in the form of socialization and individuation determined by communication in ordinary language. This interest extends to the maintenance of intersubjectivity of mutual understanding as well as to the creation of communication without domination. Technocratic consciousness makes this practical interest disappear behind the interest in the expansion of our power of technical control. Thus the reflection that the new ideology calls for must penetrate beyond the level of particular historical class interests to disclose the fundamental interests of mankind as such, engaged in the process of self-constitution. (See Habermas, 1968.)

Terminological note

'In current English, "practical" often means "down-to-earth" or "expedient". In the text, this sense of "practical" would fall under "technical". "Practical" (*praktisch*) always refers to symbolic interaction within a normative order, to ethics and politics.' (From the translator's Preface.)

References

BERGER, P. L. (1967), *The Sacred Canopy*, Doubleday.

ELLUL, J. (1967), *The Technological Society*, New York.

GEHLEN, A. (1963), 'Über Kulturelle Kristallisationen', in *Studien zur Anthropologie und Soziologie*, Berlin.

GEHLEN, A. (1964), 'Über kulturelle Evolution', in M. Holm and F. Wiedmann (eds.), *Die Philosophie und die Frage nach den Fortschritt*, Munich.

HABERMAS, J. (1967a), 'Arbeit und Interaktion: Bemerkungen zu Hegels Jenenser Realphilosophie', in H. Braun and M. Riedel (eds.), *Natur und Geschichte. Karl Lowith zum 70. Geburtstag.* (*Festschrift*), Stuttgart. (This essay is reprinted in *Technik und Wissenschaft als 'Ideologie'*, Frankfurt am Main, 1968, and will appear in English in *Theory and Practice*, to be published by Beacon Press.)

HABERMAS, J. (1967b), 'Die Klassische Lehre von der Politik in ihrem verholtnis zur Socialphilosophie', in *Theorie und Praxis*, Neuwied, 2nd edn. (To appear in *Theory and Practice*, Beacon Press.)

HABERMAS, J. (1967c), 'Naturrecht und Revolution', in *Theorie und Praxis*, Neuwied, 2nd edn.

HABERMAS, J. (1968), *Erkenntnis and Interesse*, Frankfurt am Main. (To be published as *Cognition and Human Interests* by Beacon Press.)

LENSKI, G. E. (1966), *Power and Privilege: A Theory of Social Stratification*, New York.

LÖBL, E. (1968), *Geistige Arbeit – die Wahre Quelle des Reichtums*, trans. L. Grunwald.

MACPHERSON, C. B. (1962), *The Political Theory of Possessive Individualism*, Oxford University Press.

OFFE, C. (1969), 'Politische Herrschaft und Klassenstrukturen', in G. Kress and D. Senghaas (eds.), *Politikwissenschaft*, Frankfurt am Main.

SCHELSKY, H. (1961), *Der Mensch in der Wissenschaftlichen Zivilisation*, Cologne-Opladen.

STRAUSS, L. (1963), *Natural Right and History*, University of Chicago Press.

Further Reading

Publications relevant to this field are widely diffused through the literature of both the social and the natural sciences and few specialized journals exist. There are, however, three valuable collections of papers, all of general interest:

B. Barber and W. Hirsch (eds.), *The Sociology of Science*, Free Press, 1960.
N. Kaplan (ed.), *Science and Society*, Rand McNally, 1965.
R. K. Merton, *Social Theory and Social Structure*, Free Press, 1967.

The following selection of reading is limited to works available in book form and roughly corresponds to the organization of the reader.

Part One
The Rise and Institutionalization of Science

G. Basalla (ed.), *The Rise of Modern Science*, Heath, 1968.
J. Ben-David, *The Scientists' Role in Society*, Prentice-Hall, 1971.
J. D. Bernal, *Science in History*, Watts, 1965.
H. Butterfield, *The Origins of Modern Science, 1300–1800*, Bell, 1957.
E. A. Burtt, *The Metaphysical Foundations of Modern Science*, Routledge, 1932, 2nd ed.
A. C. Crombie (ed.), *Scientific Change*, Heinemann, 1963.
A. R. Hall, *The Scientific Revolution 1500–1800*, Longman, 1962.
T. S. Kuhn, *The Copernican Revolution*, Harvard, 1957.
R. K. Merton, *Science, Technology and Society in Seventeenth-Century England*, Howard Fertig, 1970.
J. Needham, *The Grand Titration, Science and Society in East and West*, Allen & Unwin, 1969.
J. Needham, *Science and Civilization in China*, Cambridge University Press – various publication dates.
E. W. Strong, *Procedures and Metaphysics*, Hildesheim, 1966.

Part Two
Pure Science: Structural and Cultural Features

J. D. Bernal, *The Social Function of Science*, Routledge, 1939.
A. de Grazia (ed.), *The Velikovsky Affair*, University Books, 1966.
W. O. Hagstrom, *The Scientific Community*, Basic Books, 1965.
J. H. van't Hoff, *Imagination in Science*, trans. G. F. Springer, Springer Verlag, N.Y., 1967.
T. S. Kuhn, *The Structure of Scientific Revolutions*, Chicago University Press, 1970, 2nd edn.

I. Lakatos and A. Musgrave (eds.), *Criticism and the Growth of Knowledge*, Cambridge University Press, 1970.
J. T. Merz, *A History of European Thought in the Nineteenth Century*, Volumes I and II, Dover, 1965.
M. Polanyi, *Personal Knowledge*, Chicago Univ. Press, 1958.
M. Polanyi, *Knowing and Being* (ed. M. Grene), Routledge, 1969.
D. J. de S. Price, *Science since Babylon*, Yale University Press, 1961.
D. J. de S. Price, *Little Science, Big Science*, Columbia University Press, 1963.
N. Storer, *The Social System of Science*, Holt Rinehart, 1966.
R. Taton, *Reason and Chance in Scientific Discovery*, Science Editions, 1962.
J. D. Watson, *The Double Helix*, Athaneum Press, 1968.
R. M. Young, *Mind, Brain and Adaptation in the Nineteenth Century*, Oxford University Press, 1970.
J. Ziman, *Public Knowledge*, Cambridge University Press, 1968.
F. Ƶnanecki, *The Social Role of the Man of Knowledge*, Columbia University Press, 1940.

Part Three
Science Technology and Economy

J. Ben-David, *Fundamental Research and the University*, O.E.C.D., 1968.
D. S. L. Cardwell, *The Organization of Science in England*, Heinemann, 1957.
S. Cotgrove and S. Box, *Science Industry and Society*, Allen & Unwin, 1970.
B. G. Glaser, *Organisational Scientists*, Bobbs Merrill, 1964.
W. Gruber & G. Marquis (eds.), *Factors in the Transfer of Technology*, M. I. T., 1969.
J. Jewkes, et al., *The Sources of Invention*, Macmillan, 1969.
W. Kornhauser, *Scientists in Industry*, University of California, 1963.
S. Marcson, *The Scientist in American Industry*, Harper, 1960.
D. C. Pelz and F. M. Andrews, *Scientists in Organisations*, Wiley, 1966.
A. W. Warner, D. Morse and A. S. Eichner (eds.), *The Impact of Science on Technology*, Columbia University Press, 1965.

Part Four
Science and Politics

J. D. Bernal, *The Freedom of Necessity*, Routledge, 1949.
H. Brooks, *The Government of Science*, M.I.T. Press, 1968.
R. Gilpin, *U.S. Scientists and Nuclear Weapons Policy*, Princeton University Press, 1962.
R. Gilpin and C. Wright, *Scientists and National Policy Making*, Columbia University Press, 1964.
D. Greenberg, *The Politics of American Science*, Pelican, 1969.
P. Haberer, *Politics and the Community of Science*, Van Nostrand, 1968.

D. Joravsky, *Soviet Marxism and Natural Science*, Routledge 1961.
D. Joravsky, *The Lysenko Affair*, Harvard, 1970.
S. A. Lakoff (ed.), *Knowledge and Power*, Free Press, 1966.
D. Price, *Government and Science*, New York University Press, 1954.
D. Price, *The Scientific Estate*, Harvard University Press, 1965.
R. W. Reid, *Tongues of Conscience: War and the Scientists Dilemma*, Constable, 1969.
N. Vig, *Science and Technology in British Politics*, Pergamon, 1968.

Part Five
The Intelligibility and Impact of Science

F. J. Crosson (ed.), *Science and Contemporary Society*, Notre Dame, 1967.
G. Daniels (ed.), *Darwinism Comes to America*, Blaisdell, 1968.
C. C. Gillispie, *Genesis and Geology*, Harper & Row (reprint), 1959.
J. C. Greene, *The Death of Adam*, Iowa University Press, 1959.
G. Holton (ed.), *Science and Culture*, Houghton Mifflin, 1965.
S. Moscovici, *La Psychanalyse: son Image et son Public*, Presses Universitaires de France, 1961.
D. Van Tassel and M. G. Hall, *Science and Society in the United States*, Dorsey 1966.

Part Six
Scientific Concepts and the Nature of Society

J. E. Curtis and J. W. Peters (eds.), *The Sociology of Knowledge*, Duckworth, 1970
M. Douglas, *Purity and Danger*, Routledge, 1966.
M. Douglas, *Natural Symbols*, Cresset, 1970.
J. Habermas, *Wissenschaft und Technik als Ideologie*, Frankfurt, 1968 (English translation forthcoming – *Toward a Rational Society*, Heinemann, 1971).
K. Mannheim, *Ideology and Utopia* (with a preface by L. Wirth, translated by L. Wirth and E. Shils), Harcourt 1936.
H. Marcuse, *One-Dimensional Man*, Routledge & Kegan Paul, 1964.
W. Stark, *The Sociology of Knowledge*, Routledge & Kegan Paul, 1958.
T. Veblen, *The Place of Science in Modern Civilisation and other Essays*, Russell and Russell, 1961.

Acknowledgements

Permission to reproduce the Readings in this volume is acknowledged to the following sources:

1 *Science and Society*
2 *European Journal of Sociology*
3 Free Press
4 Heinemann Educational Books Ltd
5 Basic Books Inc.
6 Basic Books Inc.
7 *Social Research*
8 Ignacy Malecki and Eugeniusz Olszewski
9 MIT Press
10 OECD
11 *Technology and Society* and N. D. Ellis
12 *Science Studies*
13 *Science Studies*
14 *Daedalus*
15 S. B. Barnes
16 Dorsey Press
17 R. G. A. Dolby
18 University of Chicago Press
19 Routledge & Kegan Paul Ltd
20 Suhrkamp Verlag, Heinemann Educational Books Ltd and Beacon Press

Author Index

Subject Index

Introducing Sociology

Edited by Peter Worsley and members of the Department of Social Anthropology and Sociology, University of Manchester

Sociology is now a major area of intellectual inquiry in most countries of the world. It is also seen by an increasing number of its students as one of the most relevant of contemporary disciplines.

Introducing Sociology is an exciting and wholly original text which acknowledges both these points. It is, first and foremost, an introduction to sociological ideas and practice, not an exhaustive summary. It is written in a style which, at no sacrifice of scientific rigour, is refreshingly free from jargon. Its subject-matter is drawn from the common life-experience of most people born into the mid-twentieth century. Indeed the examples used in the book have been deliberately chosen from a wide range of cultures and societies to underline the international roots and relevance of modern sociology.

Published as a companion volume is *Modern Sociology: Introductory Readings*, prepared by the same team. A second companion volume of readings, *Problems of Modern Society: A Sociological Perspective*, will be published in March 1972.